Biochemistry Demystified

Demystified Series

Accounting Demystified
Advanced Calculus Demystified
Advanced Physics Demystified
Advanced Statistics Demystified
Algebra Demystified
Alternative Energy Demystified
Anatomy Demystified
asp.net 2.0 Demystified
Astronomy Demystified
Audio Demystified
Biochemistry Demystified
Biology Demystified
Biotechnology Demystified
Business Calculus Demystified
Business Math Demystified
Business Statistics Demystified
C++ Demystified
Calculus Demystified
Chemistry Demystified
Circuit Analysis Demystified
College Algebra Demystified
Corporate Finance Demystified
Databases Demystified
Data Structures Demystified
Diabetes Demystified
Differential Equations Demystified
Digital Electronics Demystified
Earth Science Demystified
Electricity Demystified
Electronics Demystified
Engineering Statistics Demystified
Environmental Science Demystified
Everyday Math Demystified
Fertility Demystified
Financial Planning Demystified
Forensics Demystified
French Demystified
Genetics Demystified
Geometry Demystified
German Demystified
Global Warming and Climate Change Demystified
Hedge Funds Demystified
Home Networking Demystified
Investing Demystified
Italian Demystified
Java Demystified
JavaScript Demystified
Lean Six Sigma Demystified

Linear Algebra Demystified
Macroeconomics Demystified
Management Accounting Demystified
Math Proofs Demystified
Math Word Problems Demystified
MATLAB® Demystified
Medical Billing and Coding Demystified
Medical-Surgical Nursing Demystified
Medical Terminology Demystified
Meteorology Demystified
Microbiology Demystified
Microeconomics Demystified
Nanotechnology Demystified
Nurse Management Demystified
OOP Demystified
Options Demystified
Organic Chemistry Demystified
Personal Computing Demystified
Pharmacology Demystified
Physics Demystified
Physiology Demystified
Pre-Algebra Demystified
Precalculus Demystified
Probability Demystified
Project Management Demystified
Psychology Demystified
Quality Management Demystified
Quantum Field Theory Demystified
Quantum Mechanics Demystified
Real Estate Math Demystified
Relativity Demystified
Robotics Demystified
Sales Management Demystified
Signals and Systems Demystified
Six Sigma Demystified
Spanish Demystified
sql Demystified
Statics and Dynamics Demystified
Statistics Demystified
Technical Analysis Demystified
Technical Math Demystified
Trigonometry Demystified
uml Demystified
Visual Basic 2005 Demystified
Visual C# 2005 Demystified
Vitamins and Minerals Demystified
XML Demystified

Biochemistry
Demystified

Sharon Walker
David McMahon

New York   Chicago   San Francisco   Lisbon   London
Madrid   Mexico City   Milan   New Delhi   San Juan
Seoul   Singapore   Sydney   Toronto

The McGraw·Hill Companies

Library of Congress Cataloging-in-Publication Data

Walker, Sharon, Ph. D.
    Biochemistry demystified / Sharon Walker, David McMahon.
        p. ; cm.—(McGraw-Hill "Demystified" series)
    Includes index.
    ISBN 978-0-07-149599-8 (alk. paper)
    1. Biochemistry.   I. McMahon, David (David M.)   II. Title.   III. Series.
    [DNLM: 1. Biochemistry.   QU 4 W184b   2008]
    QP514.2.W345   2008
    612′.015—dc22

                                                            2008008077

McGraw-Hill books are available at special quantity discounts to use as premiums and sales promotions, or for use in corporate training programs. To contact a special sales representative, please visit the Contact Us page at www.mhprofessional.com.

Biochemistry Demystified

1 2 3 4 5 6 7 8 9 0   FGR/FGR   0 1 3 2 1 0 9 8

ISBN 978-0-07-149599-8
MHID 0-07-149599-1

Sponsoring Editor
   Judy Bass

Acquisitions Coordinator
   Rebecca Behrens

Editing Supervisor
   David E. Fogarty

Project Manager
   Harleen Chopra, International
   Typesetting and Composition

Copy Editor
   Upendra Prasad

Proofreader
   Ragini Pandey

Indexer
   Jeff Evans

Production Supervisor
   Pamela A. Pelton

Composition
   International Typesetting and Composition

Art Director, Cover
   Jeff Weeks

ABOUT THE AUTHORS

Sharon Walker, Ph.D., is a Diplomat of the American Board of Toxicology (DABT) and has done extensive research in various areas of biomedicine. She has taught graduate courses in immunology, epidemiology, cell biology, and statistics. Dr. Walker is the author of *Biotechnology Demystified*.

David McMahon holds advanced degrees in physics and mathematics, and works as a researcher at a national laboratory. He has engaged in a wide variety of areas including fusion energy research and quantum information science. David has an interest in the biological sciences and participated for several years in research on the role of neurotransmitters in attention deficit disorder.

CONTENTS

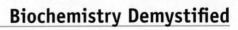

PREFACE

Biochemistry is the process of life. A knowledge of biochemistry is essential to all students of the life sciences, including biology, genetics, health, nutrition, and pathology. This book is intended to give beginning students a fundamental insight into the basic processes of biochemistry. The beauty and intricacy of biochemistry can be lost in a complex and terminology-intensive presentation. We hope to help you eat this elephant by giving you one bite at a time.

CHAPTER 1

The Chemistry of Biomolecules

Review of Basic Chemical Concepts

You, biochemist to be, have happened upon this field at a remarkable time. Human health care is undergoing revolutionary changes based on new abilities to understand and to manipulate the human system at a molecular level. You have a chance not only to understand this remarkable machine but also to remove defects and improve life. But first you need to know biochemistry.

You will need to have a background in the discipline of chemistry to understand biochemistry because biomolecules are controlled by the same forces that define other chemical processes. In this chapter we will undergo a broad review of basic tenets of chemistry. You need to understand:

- Forces that hold atoms and molecules together
- Forces that drive reactions

- The unique measurements of hydrogen ion concentration—pH
- The unique properties of the carbon atom
- The importance of molecular arrangement, or stereochemistry

Forces That Hold Atoms and Molecules Together

To understand this review, you need to know that positively-charged particles called protons exist in the nucleus of an atom, together with neutral particles called neutrons. Negatively charged electrons exist in orbitals surrounding the nucleus. Atoms have the same number of electrons as protons, so the atom is neutrally charged. Simple enough? For the purposes of this discussion, you can envision the negatively charged electrons as existing in defined orbitals, or shells, around the nucleus. This is the familiar model of the atom, called the Bohr atom (after Dr. Bohr, of course).

The electrons surrounding the nucleus exist in a specific region of space resembling clouds at specific distances from the nucleus. For the discussion of the electron orbitals, envision a series of shells that must be filled in a certain order. Note that it is actually more correct to consider the electrons as existing in a cloud around the nucleus. Some positions in the cloud are more probable than others, resulting in the simplistic picture of layers. This idealized structure is frequently compared to the skins of an onion. The electron shells actually represent energy levels. Each major shell has subshells. Each subshell can hold a maximum of two electrons. Major shells have different numbers of subshells, and therefore hold different number of electrons. Shell number one holds two electrons (one subshell). Shell number two holds eight (four subshells). Shell number three can hold up to eighteen (nine subshells) (Table 1-1).

Table 1-1   Electron Content of Major Energy Shells

Shell Number	Number of Electrons in Shell (Maximum)
1	1 subshell: 2 electrons
2	4 subshells: 8 electrons
3	9 subshells: 18 electrons
4	16 subshells: 32 electrons
5	25 subshells: 50 electrons

Here are the rules for filling shells. Subshells are filled from the lowest energy shell first, which is the shell closest to the nucleus. Electrons can be induced, by an energy input, to move to a higher shell than the lowest one available. However, they will quickly return to the more stable, lower energy configuration; and when they move to the lower energy level, they release energy. In general, atoms that don't have a full major shell of electrons will seek to either retrieve electrons from other atoms, or to give up some of their own. The electrons that will move between atoms, in the outer electron shell, are called the *valence electrons* (Fig. 1-1). In general, if the outer major shell is full, the atom is inert. Examples of inert elements are helium, neon, and argon. Inert elements have no valence electrons.

Hydrogen, for example, has one electron in the first shell that exists—the shell with the lowest possible energy state. (Remember, nature seeks the lowest energy state available.) There is one electron because hydrogen has only one proton and the number of electrons in an atom equals the number of protons. The first shell needs to have two electrons before it is full; yes, it wants one more electron than it has protons. So, hydrogen will react easily with other atoms that will share an electron. Contrast the behavior of hydrogen with that of helium. Helium has two electrons. The electron shell is perfectly "happy," and helium will not react with anything.

When the first shell is full, the next shell, with a higher energy level, receives the additional electrons. The next element is lithium, which has three protons and three electrons. The first shell has two electrons and is stable. However, that third electron in the outer shell "wants a buddy badly," so lithium is highly reactive and will give up the third electron at the drop of a hat.

There are two primary ways that atoms can associate with one another to form molecules, through *ionic* or *covalent* bonds. Both types of bonds involve the outer electron shell of the atoms.

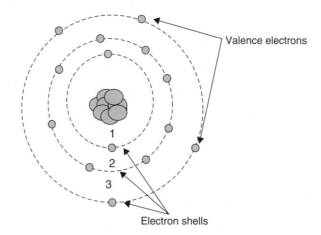

Figure 1-1   Electron energy shells and valence electrons.

IONIC AND COVALENT BONDS

Let's use hydrogen and lithium to explain the difference between ionic and covalent bonds. *Ionic bonds*, shown in Fig. 1-2, are formed when one atom gives up an electron to another atom. Ionic bonds are based on the electrical attraction of a positively charged atom to a negatively charged atom. Lithium will donate its extra third electron to another atom, such as chlorine, that is "short a full house." When the electron moves to chlorine, the lithium assumes a positive charge, and chlorine receives a negative charge by picking up one more electron than it has positively charged protons. If this reaction occurs while these atoms are in aqueous solution, the positively charged and the negatively charge chlorine will float around independently of one another. See below for a discussion of how the water molecules actually work to disperse the charged atoms.

Table salt (sodium chloride) is a familiar example of a molecule that forms ionic bonds in this fashion.

Covalent bonds, shown in Fig. 1-3, are formed when atoms share electrons; the electron actually belongs to both molecules. Let's return to our example of hydrogen. Hydrogen, as you recall, is one electron short of a full house. Two hydrogen atoms will share their electrons, forming the hydrogen gas, or H_2. The hydrogen molecule contains two protons, one per hydrogen atom, and two electrons, shared equally by these atoms. The molecule is neutral charge and has a full shell of electrons. Covalent bonds are generally more stable arrangements than ionic, especially in an aqueous environment.

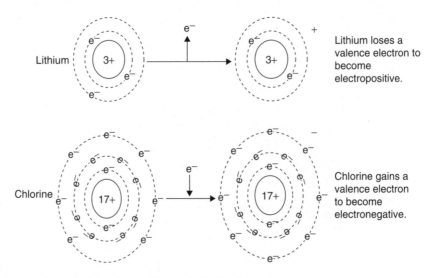

Figure 1-2   Ionic bonds.

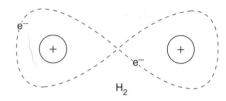

Figure 1-3   Covalent bonds.

HYDROGEN BONDS

There are other forces between molecules that affect their interactions. The general term for forces which act between stable molecules is *van der Waals* forces. They differ from covalent and ionic bonding in that they are created by momentary polarization of particles and are less stable. Remember, based on quantum theory, electrons are not fixed in the atom or molecule, but rather may exist in a number of positions as predicted by probability methods. Unlike the idealized model of onion skins, the electrons are actually not evenly distributed and may create temporary electromagnetic fields that attract or repulse one another.

Hydrogen bonds are a type of van der Waals force. These are created when one atom in a molecule has a stronger pull than the other, resulting in "polar" regions on the molecule. Consider water. Oxygen has eight electrons. Two occupy the first major shell and six occupy the second major shell. However, the second major shell can hold eight electrons, so oxygen can accept two more electrons. Water is formed when an oxygen atom picks up the two electrons from each of two hydrogen atoms. The oxygen atom shares these electrons with the hydrogen atoms. However, the sharing is not equal, as shown in Fig. 1-4.

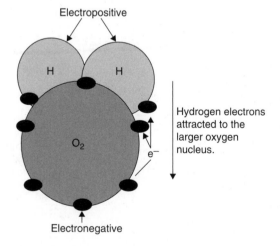

Figure 1-4   Formation of poles in a covalently bonded molecule.

Oxygen is a more electronegative element than hydrogen, meaning that it tends to pull electrons toward itself. This causes the "O" end of the H_2O molecule to be slightly negatively charged and the "H" end of the molecule to be slightly positively charged.

These partial charges at different ends of a molecule are called *dipoles*. The existence of these dipoles causes the formation of hydrogen bonds. Hydrogen bonds are due to forces between the electropositive hydrogen in one molecule and an electronegative atom in another molecule. Water is the perfect example of hydrogen bonds.

Consider two water molecules coming close together, as in Fig. 1-5.

The electropositive hydrogen is strongly attracted to the electronegative poles of the oxygen. Each water molecule can potentially form several hydrogen bonds with surrounding water molecules. Water forms a matrix as the dipoles form between the molecules. This is why water molecules appear as though they are trying to cluster together and push out anything in the way.

Hydrogen bonds form between hydrogen-containing molecules other than water. They are very important in the interactions of biomolecules. Many biomolecules literally bristle with hydrogen atoms. These atoms are electropositive and form bonds with electronegative atoms, including carbon, in other molecules. Among other things, the formation of hydrogen bonds is a primary force in the way proteins fold in on themselves. Hydrogen bonds have about a tenth of the strength of an average covalent bond. They are easily broken and reformed.

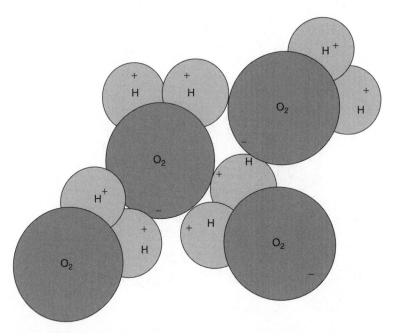

Figure 1-5   Water and the formation of hydrogen bonds.

The ionic forces within a water environment play a very important role in the behavior of biomolecules. Water tends to separate atoms or molecules that bear an electric charge. Consider our example of lithium chloride, a molecule formed by ionic bonds. In water, the ions become separate entities, and the poles on the water molecules attract positive and negative ions individually, as illustrated in Fig. 1-6.

If you keep adding lithium and chlorine, eventually there are so many atoms that the water cannot keep them apart. The opposite charges will cause these atoms to associate with one another and reform lithium chloride salt. The lithium chloride salt is a molecule; however, the force of electrical attraction that keeps these two atoms associated is not strong and can be easily disrupted by adding more water. And an atom or molecule with a stronger charge can definitely compete for one of the "partners."

Some biomolecules have surfaces charges. These are called *polar*, because they display electromagnetic poles. Such molecules move freely through an aqueous solution. On the other hand, if molecules with no ionic charge are added to an aqueous solution, the water molecules tend to cluster because they are attracted to one another and remain separate from the nonionized molecules. Biomolecules that have no surface charge, that is, are *nonpolar*, do not move freely through an aqueous solution. Rather, the nonpolar molecules tend to cluster together.

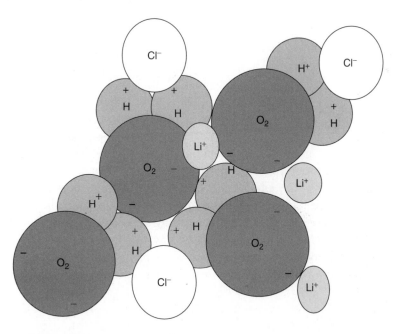

Figure 1-6   Dispersal of lithium chloride in an aqueous solution.

Forces That Drive Reactions

Our world, big and small, is controlled by the laws of thermodynamics, which is the laws of energy and heat. Change of any type requires energy, including changes in chemical systems. Thermodynamics allows us to understand and predict chemical reactions given parameters of energy exchange.

You need to master a few fairly simple concepts. The first of these is the concept of open and closed systems. A closed system is totally isolated from its environment. As far as we know, the universe is a closed system. For most purposes, the solar system is a closed system. Living systems are open systems because they interact with their surroundings, absorbing and releasing energy.

Other concepts you will need to understand the dynamics of chemical reactions are:

- Energy (U)—total energy content of a system
- Enthalpy (H)—energy plus the state functions of pressure and volume; heat content (in systems with constant pressure, volume)
- Entropy (S)—degree of randomness
- Free energy (G)—capacity to do work

These concepts are related. What you, biochemist, need, bottom line, is the value G, because the change in free energy predicts whether or not a reaction will go forward.

ENERGY

There are only two forms of energy, *kinetic* and *potential*. Kinetic energy is energy of motion. The most familiar example of kinetic energy is heat. Heat is a reflection of the rate of molecular motion in the environment. The faster the molecules move, the greater is the heat contained by the system. Potential energy is all other forms of energy, for example, chemical, electromagnetic, and positional (object with the capability of falling). One form of energy can be transformed into another. Consider the example of a man lifting a box onto a high shelf. The energy in the muscles that perform the lifting is chemical energy, a form of potential energy. Some of the chemical energy is converted into the potential energy represented by the position of the box on the high shelf. The movement of the box as it is being lifted is kinetic energy, also extracted from the chemical energy.

The first law of thermodynamics says that within a *closed system*, energy can neither be created nor destroyed. This is called the conservation of energy. By definition, a closed system does not interact with its environment and so does not exchange energy. The universe is a closed system. However, an *open system* can absorb and release energy to and from a surrounding environment. A biological system is an open system.

For biological systems, change in the energy content of the system can be measured by the heat absorbed or released. In the paragraphs that follow, you will see why energy changes can be measured so easily for you, biochemist. You will also be introduced to the concept of *enthalpy*. For biological systems, enthalpy can be conceptualized as heat content.

We represent energy in a system as U. According to the first law of thermodynamics, in a closed system, the final amount of energy (U_{final}) must equal the initial amount of energy ($U_{initial}$) (conservation of energy). However, in an open system, energy is not conserved. There can be a change in the overall energy (ΔU) because of the interaction with the surroundings.

Let:

q = heat absorbed by the system from the surroundings

w = work done by the system

ΔU = change in energy U

Heat is defined as the energy that flows between systems at different temperatures and work is defined as a force acting over distance. The change in energy in a reaction is

$$\Delta U = U_{final} - U_{initial} = q - w \qquad (1\text{-}1)$$

This tells us simply that the final energy in a system will increase from the initial energy if heat is absorbed and will decrease if work is done—which is exactly what you would predict. Consider the example of a man lifting a box. In this case, heat will be released to the surroundings (out of the system) when work is done by the muscles. In this case, q is a negative value and the final energy will decrease by the equivalent of q. Final energy will also decrease by the work done (force applied to move the box multiplied by the distance moved by the box). In systems where the q is negative (heat is given off), the reaction is called *exothermic*. If q is positive, heat is absorbed and the reaction is called *endothermic*.

Energy is considered a *state function* because it is a characteristic of a system in equilibrium. (A system is in *equilibrium* if measurable quantities are not changing even though individual components within the system may be moving around.) By contrast, neither q nor w is a state function because they both describe a process rather than a characteristic of the system.

There are many ways to measure energy. Historically, biochemists have used the *calorie*, a unit of heat, as a convenient unit of measure. One calorie is the amount of energy required to raise the temperature of 1 g of water at 14.5°C to 15.5°C. The calories used to refer to food in popular dialogue are actually *kilocalories,* where 1 kcal = 1000 cal. Of course energy in biological systems is used for functions other than generating heat. However, you can determine the total energy in a biomolecule by measuring the calories generated when the molecule is totally oxidized (burned) at a constant temperature and constant volume or pressure. This is done in an instrument called a calorimeter. The chemical energy in the molecule is converted to the kinetic energy of heat. This method of measuring the energy content of a biomolecule is possible because the total energy harvested from oxidizing the molecule is the same whether the energy is released all at once, as in the calorimeter, or in a series of steps, as in the body when a molecule is metabolized.

You will also find discussion of energy in biological systems measured in *joules*. Unlike the unit calories, joules are part of the International Systems of Units, abbreviated SI. One joule is the amount of energy required to apply 1 N of force over a distance of 1 m. One calorie is equal to 4.184 J.

ENTHALPY

Enthalpy, represented by the symbol *H*, is another way to characterize the physical state of a system. It has been described as the energy required to build a system from scratch. This state function incorporates not only energy content but also the pressure and volume of the system. Pressure and volume are also characteristics of a system, also known as state functions.

$$H = U + PV \tag{1-2}$$

where P = pressure

V = volume

A system at high pressure will have more enthalpy than the same system at low pressure, even though the total energy content may be the same.

Once again our lives are made easier by the fact that biological systems are, for all practical purposes, at constant pressure and constant volume. Note that the evolution of gases makes small changes in U. However, for us, enthalpy and energy are the same function. The energy content of a substance, determined as the heat generated by oxidizing 1 mol in a calorimeter, is a measure of the enthalpy of the substance.

Changes in enthalpy are:

$$\Delta H = \Delta U - P\Delta V \tag{1-3}$$

We argued above that, for biological systems, change in energy can be easily measured as the heat that is generated or absorbed. As changes in energy and enthalpy are the same function in our systems, clearly heat exchange also reflects changes in enthalpy.

ENTROPY

Have you heard that scientists have developed a perpetual motion machine, but the invention has been suppressed by the government and by huge oil companies? After all, in an isolated system, energy can neither be created nor destroyed according to the first law of thermodynamics. What about the full-sized automobile that can travel 100 miles on a gallon of gas? That's being suppressed as well. After you understand the second law of thermodynamics, you will know that these stories cannot be true.

According to the second law of thermodynamics, a perpetual motion machine is not possible. And the energy in a gallon of gasoline cannot be converted with 100% efficiency into movement of an automobile. The second law of thermodynamics says that the overall randomness of the system and its surroundings must increase with time. We characterize the randomness of a system by a quantity known as *entropy*. In any given reaction (meaning any change), the entropy of the system and its surroundings, combined, increases—always. So, in a closed system, a machine cannot run forever unless you put energy into the system from the outside.

To understand entropy, consider the example of the dissolution of a cube of sugar into a glass of water. A cube of sugar is highly ordered, with the molecules in association with one another according to a predictable pattern. The process of dissolution occurs spontaneously (you don't have to add energy) and the system goes from orderly (sugar molecules in a cube) to total randomness (sugar molecules dispersed throughout the system). The entropy of the system is increased.

The requirement for randomness to increase is predictable by the laws of statistics. In the above example, it is possible that the sugar molecules could spontaneously reform a cube. However, the probability that a given molecule will appear in the cube formation as compared to the probability that this same molecule will disperse randomly somewhere in the system is really, really small, so small as to be essentially nil.

To induce the sugar to recrystallize into a cube or into another orderly arrangement, some input of energy would be required, such as adding heat to vaporize the water. Even though the order within the collection of sugar molecules will increase, that is, system entropy decreases, the overall randomness of the system plus that of the surroundings must increase—overall entropy increases. If the water is vaporized, the randomness of the water molecules' configuration is increased and temperature of the system is raised. Increased temperature reflects increased randomness in the system. The combination of the change in the configuration of the sugar molecules, the change in the configuration of the water molecules, and the increased temperature must result in an overall increase in entropy.

You may realize that, if entropy is increasing in our overall system (the universe), the universe must be running down! Over the entire universe, energy is converting to a more random form and is less able to do work. Actually, this is indeed the case, a phenomenon called "entropy doom." I suggest that you do not dwell on this.

By convention, entropy is represented by the symbol S. Change in entropy is related to the heat of reaction. For biological systems which are at constant temperature, this relationship can be described as:

$$\Delta S_{surroundings} = -\Delta H_{system}/T \tag{1-4}$$

where $\Delta S_{surroundings}$ = change in entropy

ΔH_{system} = heat transfer into the system

T = absolute temperature of the surroundings at which the transfer occurs

H is measured in calories or joules. T is in kelvins. The unit of measure for entropy is joules per kelvin (J/K).

This relationship, true only of systems in equilibrium, tells us that the overall change in entropy is inversely proportional to the amount of heat generated by the system and to the temperature of the surroundings where the transfer occurs. A system generating heat loses enthalpy and increases the entropy of the surroundings. If the heat is transmitted into a low temperature surroundings, the increase in entropy is greater than if the heat is transmitted into a high-temperature surroundings.

FREE ENERGY

At this point, you have been introduced to energy, enthalpy, and entropy. Now we are going to derive a concept called *free energy* which is a measure of the capacity of a chemical system to do work. The work can be manifested by rearranging atoms, building molecules, and in transport mechanisms.

By convention, free energy is represented by the symbol G after the worthy individual who developed this concept in the 1870s—the American mathematical physicist Willard Gibbs. The Gibbs free energy is given by

$$G = H + PV - TS \tag{1-5}$$

And at constant volume and constant pressure, and at a fixed temperature

$$\Delta G = \Delta H - T\Delta S \tag{1-6}$$

The capacity of a system to do work—the free energy—is its overall energy content minus a combination of the degree of randomness in the system and the temperature. If you think about this for a moment you will be able to understand why maintaining the system at high temperature takes some of the energy content, and why a high degree of randomness decreases the ability of any system to focus energy and do work.

What you, biochemist, need out of all of this is a method to predict which reactions will go forward spontaneously. By definition, spontaneous reactions don't require energy from outside the system. The change in free energy from the starting state to the ending state must be less than zero.

$$\Delta G < 0$$

You cannot have an increase in free energy in your end product relative to what you start with unless you bring in energy from outside the system.

Notice that if entropy decreases, S will be negative. Therefore, $-T\Delta S$ will be positive and the change in free energy will be more positive. As you create order, you create more free energy. Also, you would expect that reactions that result in a decrease in randomness of the system would be more difficult to push forward.

- Reactions where ΔH (change in enthalpy) is negative and ΔS (change in entropy) is positive will go forward spontaneously.

- Reactions wherein ΔH is positive and ΔS is negative will not go forward without energy input from the outside.

EXAMPLE

Most energy in biological systems comes from oxidation of glucose according to the following reaction:

$$C_6H_{12}O_6 + 6O_2 = CO_2 + 6H_2O$$

This reaction results in a decrease in enthalpy and an increase in entropy.

$$\Delta H = -2.808 \text{ kJ/mol}$$
$$\Delta S = 182.4 \text{ J/K/mol}$$

The reaction proceeds spontaneously. The overall change in free energy is:

$$\Delta G = \Delta H - T\Delta S = -2.808 - (310)(0.1824) = -2.865 \text{ kJ/mol}$$

The rub comes if both ΔH and ΔS are negative or both are positive.

Positive ΔH (increase in energy content) can be overcome if the increase in entropy is large enough. Decrease in entropy ($-\Delta S$) can be overcome if the decrease in energy content is large enough.

A determining factor is the temperature of reaction. Manipulation of Eq. (1-6) reveals that $\Delta G = 0$ if

$$T = \Delta H/\Delta S$$

If both enthalpy and entropy are negative, the energy of the product is decreasing but the degree of order is increasing. ΔG (change in free energy) will be negative (the reaction will go forward) if T is less than $\Delta H/\Delta S$.

If both enthalpy and entropy are positive, randomness is increasing but the energy of the product is increasing. ΔG will be negative if T is greater than $\Delta H/\Delta S$.

EXAMPLE
Consider a reaction where

$$\Delta H = 2400 \text{ kJ/mol} \quad \text{and} \quad \Delta S = 7.00 \text{ kJ/mol}$$

What is the minimum temperature at which this reaction will proceed spontaneously?

$$T > \Delta H/\Delta S$$
$$T > 2400/7.00$$
$$T > 342 \text{ K}$$

A reaction that results in a negative change in free energy is called *exergonic*. This is good. This reaction will go forward spontaneously. A process that results in a positive change in free energy requires energy from outside the system and is called *endergonic*. Not necessarily bad—just more trouble.

Activation Energy

The change in free energy reflects a difference between the beginning and end state of atomic arrangements. However, reactions that go forward spontaneously (exergonic) may proceed extremely slowly unless the *activation energy* is provided. The activation energy is the energy required for a chemical reaction to reach equilibrium quickly. Consider the ignition of wood in a fireplace. The oxidation of the wood is thermodynamically favored and proceeds at a small rate, even at room temperature. However, the reaction does not achieve combustion until the wood reaches a certain temperature. The system must be supplied with the activation energy before the reaction proceeds at a discernable rate.

The heat of a given solution is really an average of the heat embedded in individual molecules in the form of kinetic energy. Molecules with high kinetic energy are more likely to form chemical bonds than those with low kinetic energy. At intermediate temperatures, some of the molecules might be able to form the required chemical bonds, but many will not. The activation energy increases the number of molecules at the kinetic energy level needed for bond formation. Enzymes accelerate the rate of biological reactions by lowering the activation energy. They do so by physically positioning the reactants in a configuration favorable to interaction.

Equilibrium

Here is the reality. Reactions rarely go to completion. That is, the beginning reactants are rarely completely transformed into the product. Rather, the reaction obtains equilibrium. Some of the product goes backwards to re-form the beginning reactants. This phenomenon can be represented thusly:

$$aA + bB \Leftrightarrow cC + dD \qquad (1\text{-}7)$$

where A = reactant.

a = number of molecules required of reactant A.

B = reactant.

b = number of molecules required of reactant B.

C = product.

c = number of C molecules formed from a molecules of A and b molecules of B.

D = product.

d = number of D molecules formed from a molecules of A and b molecules of B.

Now, assuming that C and D represent products you desire, you, biochemist, are happier if the equation favors the right side. You can characterize the tendency of a reaction to favor either side by a quality called the equilibrium constant [K_{eq}].

$$K_{eq} = [(Cc)(Dd)]/[(Aa)(Bb)] \qquad (1\text{-}8)$$

If K_{eq} is large, most of the reactants will form products; very little reactant will remain in the system.

For example, consider the dissociation of phosphoric acid to release a hydrogen ion.

$$H_3PO_4 \Leftrightarrow H_2PO_4^- + H^+ \qquad (1\text{-}9)$$

The K_{eq} for this reaction is:

$$K_{eq} = [(H^+)(H_2PO_4^-)]/H_3PO_4 = 7.5E - 3 \qquad (1\text{-}10)$$

At equilibrium, some of the phosphoric acid dissociates, but most is found as the initial reactant.

Phosphoric acid actually undergoes a series of reactions as follows:

$$H_3PO_4 \Leftrightarrow H_2PO_4^- + H^+ \Leftrightarrow HPO_4^{2-} + H^+ \Leftrightarrow PO_4^{3-} + H^+ \qquad (1\text{-}11)$$

Each of these steps has its own equilibrium constant. When the system is in equilibrium, all of the molecules will be in existence at some level. We will return to a discussion of phosphoric acid when we discuss buffers.

A really important equilibrium for biological systems is exhibited by water. Most water is good old H_2O. However, a small fraction of water is in equilibrium, as follows:

$$H_2O \Leftrightarrow H^+ + OH^- \qquad (1\text{-}12)$$

$$K_{eq} = [H^+][OH^-]/[H_2O] = 10^{-14} \qquad (1\text{-}13)$$

The dissociation, as you can see, is quite small. Out of every mole of water, 10^{-14} th of a mole dissociates.

Free Energy Changes and Kinetics of Chemical Reactions

In the section "Free Energy," you learned that reactions will not go forward unless the change in free energy (ΔG) is negative, or downhill. In the section "Equilibrium," you learned that the relative tendency for a reaction to go to the right (product formation) versus to the left (reformation of reactants) is reflected in the equilibrium constant (K_{eq}). It is reasonable that the K_{eq} is related to the change in free energy. Large equilibrium constants are associated with reactions involving a large negative change in free energy.

The relationship between change in free energy and the equilibrium constant is as follows:

$$\Delta G = -RT \ln K_e$$

and

$$\ln K_e = \Delta G/RT$$

EXAMPLE

The enzyme phosphoglyceromutase catalyzes an intermediate step in glycolysis that converts two molecules of 3-phosphoglyceric acid ($C_3H_5O_4P_1$) to two molecules of 2-phosphoglyceric acid ($C_3H_5O_4P_1$). The phosphorus atom simply shifts position on the molecule. The K_{eq} for this reaction at physiological pH (7.4) and temperature (310 K) has been reported to be 5.0. What is the change in free energy?

$$\Delta G = -RT \ln K_e$$
$$\Delta G = -(8.31 \text{ J/K/mol})(310 \text{ K})(1.61) = -4147.5 \text{ J/mol}$$

Now we are intellectually equipped to answer a question all new biochemists have—"Why is it so complicated?" Why are the pathways for production (and destruction/metabolism) of these molecules so long and involved? Why do so many processes have multiple steps, with intermediates that I, a biochemistry student, have to memorize? The answer lies, in part, in the constraints applied by free energy.

Biological systems need to perform processes that require a gain in free energy. The way to do this is to *couple* the reaction with a process that releases free energy. The coupling can occur if the reactions share a common intermediate.

$$A + B \Leftrightarrow C + D \ (+\Delta G)$$
$$D + E \Leftrightarrow F + G \ (-\Delta G)$$

Result:

$$A + B + E \Leftrightarrow F + G \ (-\Delta G) \tag{1-14}$$

In many cases the coupled reaction involves the breaking of phosphate bonds, a process that we will discuss in detail in later chapters. Briefly, the reaction that releases a phosphate reflects a large negative change in free energy and can drive other reactions that require a positive change in free energy. The phosphate bonds are frequently described as high-energy bonds and in essence store energy for use later.

EXAMPLE

ΔG for the hydrolysis of adenosine triphosphate (ATP) is −31 kJ/mol.

$$ATP + H_2O = ADP + Pi$$

This reaction is, therefore, highly exergonic. The hydrolysis of ATP is frequently coupled with thermodynamically unfavored reactions. The phosphorylation of glucose is required for entry into the Krebs cycle. ΔG for this reaction is 16.72 kJ/mol. How many reactions involving the hydrolysis of ATP need to be coupled with the phosphorylation of glucose for the reaction to proceed? The answer is that a single ATP hydrolysis will be enough. The combined change in free energy is 16.72 kJ − 31 kJ = −14.28 kJ. Any ΔG value less than 0 indicates a spontaneous reaction.

The Unique Measurements of Hydrogen Ion Concentration—pH

Biological systems are very sensitive to the concentration of hydrogen ions. Compounds that will release a proton (note that a hydrogen atom is a proton) to an aqueous solution are called *acids*. Acids are symbolized HX, wherein, in solution, H^+ and X^- are released. In such a system, the HX molecule is called the *conjugate acid*, because it donates a proton, and the X^- is called the *conjugate base*, because it is capable of accepting a proton. Compounds that release proton acceptors into an aqueous solution are called *bases*, symbolized XOH. The OH^-, in water, binds an H^+ and causes a decrease in hydrogen ions in the system. In basic systems, X^+ is the conjugate acid and XOH^- is the conjugate base.

To better describe the concentration of hydrogen atoms in an aqueous system, biochemists (like you) have derived the concept of pH. Simply put:

$$pH = -\log [H^+] \qquad (1\text{-}15)$$

Water has a concentration of hydrogen ions of 10^{-7} moles per liter, and the pH of water is 7. A pH of 7 is neutral. Consider the pH of a 0.1 molar solution of HCl, hydrochloric acid. This acid dissociates nearly completely into H^+ and Cl^-, generating a hydrogen ion concentration of [0.1]. The log of 0.1 is −1, so the pH of this solution is 1. Acid solutions have a higher concentration of H^+ than water. Therefore, the pH is lower than 7 (because it is the negative logarithm). Basic solutions have pH greater than 7.

Note that pH is a logarithmic system. An increase in pH of 1 unit corresponds to a 10-fold decrease in hydrogen ion concentration.

BUFFERS

Buffers are solutions of molecules that, in certain pH ranges, will pick up H^+ ions at the acid end of the range and will release H^+ ions at the base end of the range. Buffers prevent radical changes in pH with the addition of either acids or bases to a solution.

The equilibrium constant for an acid, symbolized HA is determined by the hydrogen ion concentration [H⁺]. You know, therefore, that there is a relationship between the equilibrium constant (K_{eq}) and pH. Just for grins, let's convert the formulation for the equilibrium constant to logarithms.

K_a is the equilibrium constant for an acid.

$$K_a = [H^+][A^-]/[HA] \tag{1-16}$$

where $[H^+]$ = hydrogen ion concentration

$[A^-]$ = conjugate base concentration

$[HA]$ = conjugate acid concentration

$$\log K_a = \log([H^+]) + \log([A^-]/[HA]) \tag{1-17}$$

$$-\log K_a = -\log([H^+]) - \log([A^-]/[HA]) \tag{1-18}$$

Let $-\log K_a = pK_a$ (1-19)

$pH = pK_a + \log([A^-]/[HA])$—the *Henderson-Hasselbalch* equation   (1-20)

Now we come to the crux of the matter. For $[A^-]/[HA]$ to be equal to 1, the acid must be 50% dissociated. Also, if $[A^-]/[HA]$ is equal to 1, pH is equal to the pK_a. Buffers are most effective at the 50% dissociation mark—or when pH is equal to pK_a. If you know the equilibrium constant, you know pK_a and can predict where (pH range) the buffer will be most effective.

Most of the buffering capacity of biological fluids is provided by phosphates and carbonates. Let us return to the three-step dissociation of phosphoric acid. The pK_a of the first step is 2.12, of the second is 7.21, and of the third is 12.67.

$$H_3PO_4 \Leftrightarrow H_2PO_4^- + H^+ \Leftrightarrow HPO_4^{2-} + H^+ \Leftrightarrow PO_4^{3-} + H^+ \tag{1-21}$$
$$\quad 2.12 \qquad\qquad 7.21 \qquad\qquad 12.67$$

Most biological processes occur at near-neutral pH (7.4). Phosphoric acid is able to act as a very important buffer in biological systems because at pH 7.4, the phosphate exists as a nearly equal mixture of $H_2PO_4^-$ and HPO_4^{2-}. If hydrogen ions (H^+) are added to the solution, the excess hydrogen combines with HPO_4^{2-} to form $H_2PO_4^-$. If hydroxyl ions (OH^-) are added to the solution, the $H_2PO_4^-$ releases H^+ to combine with the OH^- and form H_2O and HPO_4^{3-}. In either case, the pH remains stable.

EXAMPLE

Consider a system at pH 7.4 (physiological pH). For simplicity sake, assume that the only buffer in the system is phosphoric acid and ignore all but the dissociation constant near 7.4, that is, 7.21. At this pH, the other hydrogen ions on phosphoric acid will play a relatively minor role in regulating pH.

$$pH = pK_a + \log([A^-]/[HA])$$

$$A^- = HPO_4^{2-}$$

$$HA = H_2PO_4^-$$

$$7.4 = 7.2 + \log([A^-]/[HA])$$

$$\log([A^-]/[HA]) = 0.2$$

$$[A^-]/[HA] = 1.6$$

In this particular system,

$$[H_2PO_4^-] = 0.01 \text{ mol/L}$$

$$[HPO_4^{2-}] = 0.016 \text{ mol/L}$$

What is the change in pH resulting from the addition of 0.001 M of NaOH?
The $[A^-]/[HA]$ becomes 0.0161/0.009 because the OH^- ion adds 0.001 mol/L to the base term (A^-) and causes dissociation of 0.001 mol/L of the acid term (HA).

$$pH = 7.2 + \log([0.0161]/[0.009]) = 7.2 + \log 1.8 = 7.2 + 0.26 = 7.46$$

The addition of the base creates a relatively minor change in pH.
By contrast, the addition of 0.001 mol/L of NaOH to pure water results in the following pH change:

$$H^+ = 10^{-14}/0.001 = 10^{-11}$$

$$pH = 11$$

The Unique Properties of the Carbon Atom

Life is based on the carbon atom. Certainly there are other atoms (such as silicon) that could play the role that carbon plays as a structural backbone. However, there is no other material that has both of the key properties of carbon:

- Ability to accept four valence electrons, allowing simultaneous association of a carbon atom with four other atoms
- Highly exothermic oxidation reaction—oxidation of reduced carbon releases much energy.

VALENCE STRUCTURE OF CARBON

Carbon has six electrons.

- Major electron energy shell number one is full with two electrons.
- Shell number two has *four* electrons and can hold eight electrons—it lacks four electrons.

To obtain these four valence electrons, carbon will form covalent bonds with hydrogen, oxygen, sulfur, and nitrogen and other carbon atoms, allowing the formation of the biomolecules that constitute life.

Carbon will also form a single or a double bond with other carbon atoms (or oxygen or nitrogen). Carbon atoms will form a ring where each carbon atom shares one or two electrons. If the ring behaves as though the carbon atoms were connected by alternating single and double bonds, the compound is called *aromatic*. In the diagrams shown in Fig. 1-5, the sharing of two electrons is depicted by a double line, whereas the sharing of only one electron is depicted by a single line. The benzene ring, shown in Fig. 1-5 is the simplest aromatic compound. Compounds where all the carbon atoms are linked by a single bond and hold two hydrogen atoms as well are called *saturated*. In our unfolding discussion of biochemistry, we will see that many molecules exist as a straight chain of carbon in equilibrium with a ring structure. Fig. 1-7 is a schematic of the carbon atom together with a diagram of several simple compounds formed from carbon.

THE ROLE OF CARBON IN THE CELL ENERGY CYCLE

The energy for life comes as a result of a chemical reaction that releases energy based on the exchange of electrons between molecules. Specifically, the energy for the overwhelming majority of life on earth comes from the *oxidation* of a reduced form of carbon. As luck would have it, this particular reaction releases an unusually large amount of energy. But you know this! You oxidize a reduced form of carbon in your fireplace to create heat! Let us explain.

First, let's review what *oxidation/reduction* means. Oxidation is named after oxygen. Substances that accept electrons have the ability to oxidize other substances and are said to be *oxidative* and are known as *oxidizing agents, oxidants,* or *oxidizers*. Oxygen is a very good oxidizer because it is "electron-hungry" and will remove electrons from another substance. Substances that have the ability to *reduce* other substances are said to be *reductive* and are known as *reductive agents, reductants,* or *reducers*. The reductant transfers electrons to the other substance. This is very difficult to keep straight because ultimately, the reductant will have been oxidized and the oxidant will have been reduced. The important thing to remember for our purposes is that oxygen will oxidize carbonaceous material. The most complete

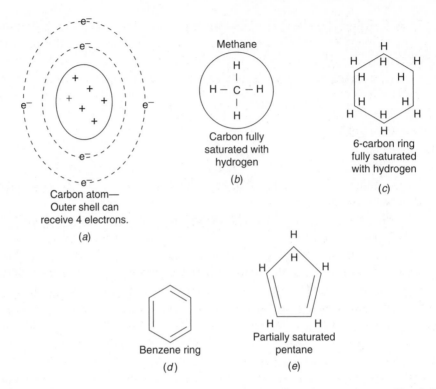

Figure 1-7 The wonderful carbon atom showing (*a*) the valence structure (*b*) a fully saturated carbon atom (saturated with hydrogen), forming methane, (*c*) a 6-carbon molecule forming a ring where each carbon is linked to another, (*d*) a benzene ring where each carbon is double bonded to another, and (*e*) a 5-carbon molecule forming a ring where some of the carbons are double bonded.

reaction results in the formation of CO_2 and H_2O. In your fireplace, the carbon in the wood is found in molecules containing a large hydrogen content, which means that the carbon is taking electrons from the hydrogen. This is a reduced form of carbon. When the oxygen in the air oxidizes this carbon into carbon dioxide and water, the electrons are transferred from the carbon-hydrogen bond to the carbon-oxygen and hydrogen-oxygen bonds and you benefit from the large energy release.

$$C_xH_x + xO_2 = xCO_2 + xH_2O$$

Here is what happens at a molecular level. Remember in the discussion of the structure of an atom, you learned that electrons exhibit energy levels. Electrons seek the lowest available energy level, nearest the nucleus. However, they can be induced by energy input to move to a higher shell. When they return to the lower energy shell, as they inevitably do, energy is released. Energy is released from the oxidation of reduced carbon, because when the electrons are transferred from hydrogen to oxygen, they move to a lower energy level. Hydrogen and carbon share the electron equally. Neither atom is particularly electronegative, so the electron is balanced fairly evenly between the nuclei of the two atoms. However, oxygen is strongly electronegative and pulls an electron shared either with carbon or with hydrogen closer to the oxygen nucleus. As the electron moves closer to the oxygen nucleus, away from the carbon or the hydrogen nucleus, the electron loses energy to the surroundings.

The Importance of Molecular Arrangement—Stereochemistry

Imagine that you need to describe a socket wrench to an alien being from another planet. You might start with the material—carbon steel, aluminum, and so on. You might even include some details, such as whether or not the tool is galvanized, or contains zinc. The material is important because it gives the tool strength. But the real issue is the shape. And not just the general shape, but the exact size and shape of the opening of the wrench. If you need to actually apply the tool, the shape must be specified exactly, depending on the specific application. As is true with a socket wrench, the use of biomolecules depends as much on their shape as their composition. Allow us to illustrate.

Consider the sugars glucose, galactose, and mannose. Glucose is the main product of photosynthesis and is the source of energy for the cell. The formula for glucose is $C_6H_{12}O_6$. Galactose is a sugar found in dairy products and in sugar beets. It binds with glucose to form lactose, the sugar found in milk. Mannose is a sugar found in some fruits, including cranberries. However, both mannose and galactose must be converted to glucose before they can be used for energy. The formula for galactose is $C_6H_{12}O_6$. The formula for mannose is $C_6H_{12}O_6$. All three sugars have the same composition. They are *isomers* of one another. The difference is in the way the atoms are arranged. To visualize the three-dimensional arrangement of atoms in a molecule using a two- dimensional piece of paper, we have to use a trick or two. To continue our example of the three sugars, glucose, galactose, and mannose, we will use *Fischer* diagrams. Fischer diagrams are constructed by the following rules: bonds that are oriented vertically are assumed to project away from the viewer and bonds that are oriented horizontally are assumed to project toward the viewer. Here are the Fischer diagrams for the three sugars (Fig. 1-8).

By convention, the carbon with the *carbonyl group* is considered carbon number 1. A carbonyl group is a functional group composed of a carbon atom double-bonded to an oxygen atom: C=O. Mannose differs from glucose by the orientation of the —OH (hydroxyl group) and the —H around carbon number 2. Galactose differs

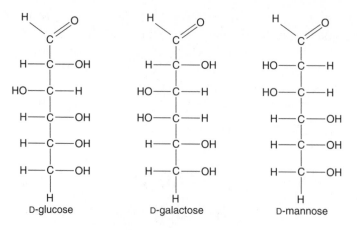

Figure 1-8 Glucose, galactose, and mannose.

from glucose by the orientation of the —OH and the —H around carbon number 4. And that makes all the difference. The D designation means that the molecule will rotate light in a clockwise direction and is *dextrorotatory*, as opposed to *levorotatory*, rotating light in a counterclockwise direction. These concepts are further explained in Chap. 3.

WAYS TO REPRESENT THREE-DIMENSIONAL MOLECULES

In our discussion of sugars, we looked at Fischer diagrams. However, Fischer diagrams quickly become unreadable for complex molecules. Also, the rules in constructing these diagrams are not intuitive. There are other methods to represent molecular structures which are more realistic.

One convention represents molecules that project toward the reader as solid arrows with the thick end closest to the viewer. Bonds that project away from the reader are striped or shaded arrows with the thick end furthest from the reader. These representations are more intuitive. See Fig. 1-9.

Our example sugars have been thus far shown in their linear formation. This formation exists in equilibrium with a ring structure that is represented with a Hawthorne projection. See Fig. 1-10.

Another representation of three-dimension structures shows the bonds between molecules as sticks and the molecules as little knobs on a stick. Below is depicted the ball-and-stick representation of D-glucose. By convention, oxygen molecules are red, hydrogen is blue, and carbon is black. In a black-and-white text like this one, oxygen will be white, hydrogen grey, and carbon black. This representation gives you a feel for the relative positions of the atoms within molecules. See Fig. 1-11.

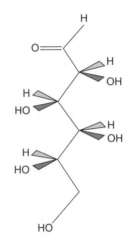

Figure 1-9   D-mannose.

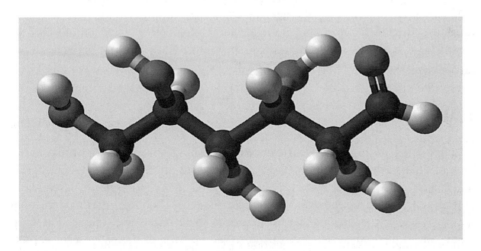

Figure 1-10   Hawthorne projection of D-glucose.

Figure 1-11   D-glucose.

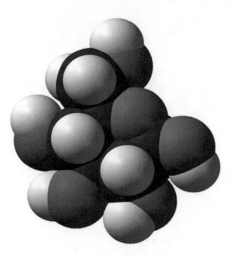

Figure 1-12　D-glucose: another view.

However, this depiction is a distortion—the bonds are much too large. An actual molecule is more like a collection of overlapping spheres. The molecules exert a sphere of influence upon one another, as they share electrons. The result, as in Fig. 1-12 above, looks somewhat like soap bubbles, piling upon one another.

These depictions show the physical conformation of complex molecules, and allow viewing of the arrangement of binding sites, and so on. With very large molecules, the visual impact is overwhelming. While such depictions are impressive, I find that this depiction conveys a limited amount of information, especially with readers new to the subject. The viewer registers two things—big and complex. For most of our needs, the Fischer diagrams or similar strategies will suffice. But remember that these are a gross oversimplification of the physical world. Remember that biomolecules are mostly large, complex, and form intricate shapes that allow them to function.

Summary

Living beings are biosystems. Biosystems (meaning you) build and break down biomolecules. The reactions for these activities are governed by the same forces that govern all chemical systems. To build molecules, atoms form associations through ionic, covalent, and hydrogen bonds. All of these bonds are due to interactions between the electrons in the outermost electron shells of the atom in question. Ionic bonds are formed when one atom acquires an electron at the expense of another atom. When this happens, the atom that picks up an extra electron assumes a negative charge and the atom that gives up an electron assumes a positive charge. The positive atom and the negative atom experience an electromagnetic attraction to one another. This attraction holds the molecule together. Sodium chloride (table salt) is an example of a molecule held together by ionic bonds. Covalent bonds are formed when atoms share electrons. Acetic acid (vinegar) is an example of a molecule formed through covalent bonds. Hydrogen bonds are formed when atoms within a covalently bonded molecule share electrons unevenly. The electron binds more closely to one atomic nucleus than to the other, creating positive and negative dipoles. For example, in a water molecule (H_2O), the oxygen atom nucleus holds the two electrons donated by the hydrogen atoms more closely than do the hydrogen atom nuclei. As a result, the oxygen area of the molecule is electronegative and the hydrogen areas of the molecules are electropositive. The electronegative molecular poles of one molecule will attract the electropositive molecular poles of other molecules.

Forming and breaking down molecules requires energy. Each molecule has a measurable energy content. The energy content (U) of a molecule can be characterized by any of the interchangeable units of energy. In a biomolecule, energy content is synonymous with heat content (H or enthalpy). Energy content of biomolecules is typically measured in calories, determined by the heat released when the molecule is completely oxidized (or burned). Each biochemical reaction can be viewed as a system. According to the first law of thermodynamics, in any closed system, energy can neither be created nor destroyed. However, biological reactions are not closed systems, because they exchange heat and material with their surroundings. For biological systems, energy can be infused from the environment.

The second law of thermodynamics says that the overall randomness of the system and its surroundings must increase. The characteristic of randomness is called entropy (S). In any given reaction (meaning any change), the entropy of the system and its surroundings, combined, increases *always*.

In any system, some of the energy is tied up in maintaining the system at a certain temperature and in maintaining a certain degree of order in the system. The remaining energy is available to do work. This "available energy" is known as free energy and is represented by the symbol G for Gibbs free energy. For chemical systems, the free energy content is also known as chemical potential. For a chemical reaction to proceed spontaneously, the reaction must result in a loss of free energy ($-\Delta G$). In other words, the products must have less free energy than the reactants. Bioreactions frequently couple reactions with a positive change in free energy with reactions with a larger negative change in free energy through a common intermediate.

Given time, all reactions establish an equilibrium where product is forming and dissociating into reactants at a predictable rate. A measure of the status of a chemical system in equilibrium is the equilibrium constant (K_{eq}). For a reaction in equilibrium, K_{eq} is the ratio between the product of the concentration of the molecules on the right side of an equation (the products) and the product of the concentrations of the molecules on the left side of the equation (the reactants).

Bioreactions occur in aqueous solutions where the hydrogen concentration is a critical parameter. Hydrogen ion concentration is measured by pH, which is the negative log of the H^+ concentration. Water has a pH of 7. Because the parameter is expressed as a negative, solutions with fewer free hydrogen ions than water have a pH greater than 7 and are basic. Solutions with more free hydrogen ions than water have a pH less than 7 and are acidic. Buffers are molecules that will release a hydrogen atom in a basic environment and will absorb a hydrogen ion in an acidic environment. Biosystems are exquisitely sensitive to pH and employ buffers, primarily carbonic acid and phosphoric acid, to minimize pH changes.

Living systems are built on carbon. Two qualities of carbon position this molecule to form the backbone of all biomolecules and to serve as the energy source that drives life. Carbon seeks four electrons to complete the outer electron energy shell. Carbon therefore can pick up these electrons from four different atoms or two carbons can share a pair of electrons and each find electrons from two other atoms. A carbon atom can form a double bond with oxygen and nitrogen and with other carbon atoms. The enormous diversity of life is provided by the versatility of the carbon atom's bonding. Oxidation of a reduced form of carbon provides energy for most biosystems. When carbon goes from a reduced state (usually bonded with hydrogen) to an oxidized state (bonded with oxygen), a large amount of energy is released, fueling the processes of life.

The function of a biomolecule depends on the arrangement of the atoms in the molecule. Molecules with the same chemical formula but different arrangements of the atoms (isomers) can have quite different functions. One of the challenges for anyone so arrogant as to try to put a complex, three-dimensional biomolecule on a two-dimensional sheet of paper is to fairly represent such a molecule. Fischer diagrams are one methodology using lines to represent bonds, whereby vertical bonds project away from the reader and horizontal bonds toward the reader. Fischer diagrams are useful to present the basic structure of a molecule. The Fischer system is not intuitively obvious, however, and quickly becomes complex. Other methods use thickening of a linear bond representation if the atom attached to the bond projects away from the reader and thinning of a linear bond representation if it projects toward the reader. This scheme allows the actual structure of a molecule to be more readily visualized. Other representations show the atoms as spheres, representing the sphere of influence. Such depictions are more accurate, but become extremely complex. They do, however, show the physical conformation of a complex molecule, allowing views of binding sites, for example.

Quiz

1. Many biosystems require hydrogen bonds. In other words,

    (a) the hydrogen gives up an electron to another molecule.

    (b) hydrogen shares an electron with another molecule.

    (c) hydrogen shifts between molecules.

    (d) the electronegative pole of one molecule attracts the electropositive pole of another.

2. A reaction will occur spontaneously if

    (a) there is an increase in entropy.

    (b) there is an increase in free energy.

    (c) there is a decrease in free energy.

    (d) heat is absorbed.

3. A decrease in entropy in a reaction (a + ΔS_{system}) can be overcome by

    (a) an increase in entropy in the surroundings ($-\Delta S_{surrounding}$).

    (b) if the temperature of the reaction is lower than the ratio of $\Delta H / \Delta S$.

    (c) if the temperature of the reaction is greater than the ratio of $\Delta H / \Delta S$.

    (d) Both (a) and (b) are correct.

    (e) Both (a) and (c) are correct.

4. A large equilibrium constant (K_{eq}) means

    (a) the system in equilibrium consists mainly of product.

    (b) the system in equilibrium consists mainly of reactant.

    (c) the reaction will not go forward spontaneously.

    (d) the reaction will absorb heat.

5. A solution that has a pH of 5

    (a) has twice the hydrogen ion concentration of a solution with a pH of 10.

    (b) has half the hydrogen ion concentration of a solution with a pH of 10.

    (c) is acidic.

    (d) is basic.

    (e) Both (a) and (c) are correct.

    (f) Both (b) and (d) are correct.

6. A solution that has a pH of 1

   (a) has twice the hydrogen ion concentration of a solution with a pH of 2.

   (b) has half the hydrogen ion concentration of a solution with a pH of 2.

   (c) has ten times the hydrogen ion concentration of a solution with a pH of 2.

   (d) is a buffer.

7. Entropy

   (a) is a state function of a system that describes the degree of order.

   (b) always increases in a closed system.

   (c) always decreases in a closed system.

   (d) can neither be created nor destroyed.

   (e) Both (a) and (b) are correct.

   (f) Both (a) and (c) are correct.

8. Buffers

   (a) release a hydrogen ion when pH drops.

   (b) release a hydrogen ion when pH rises.

   (c) absorb a hydrogen ion when pH drops.

   (d) absorb a hydrogen ion when pH rises.

   (e) Both (b) and (c) are correct.

   (f) Both (a) and (d) are correct.

9. Carbon

   (a) absorbs energy when it combines with oxygen.

   (b) possesses a large amount of free energy in a reduced form.

   (c) releases energy when oxidized.

   (d) releases free energy when reduced.

   (e) Both (a) and (d) are correct.

   (f) Both (b) and (c) are correct.

10. L and D forms of a molecule

   (a) are isomers.

   (b) rotate light in different directions.

   (c) have atoms arranged differently around all carbons.

   (d) are identical.

   (e) Both (a) and (b) are correct.

CHAPTER 2

Cell Structures and Cell Division

Introduction

Most of us are struck by the tremendous diversity of life forms. From our vantage point, the world is a display of different solutions to the problem of life. In some cases, life is vested in bizarre creatures with whom we would appear to share little in common. Sea anemones come to mind. This view is very far from the truth. All living creatures operate with the same tools, that is, the same or very similar chemicals. The commonalities of all organisms bespeak a close relationship—we share so many of the same molecules, the same strategies for fundamental processes and the same fate. We are more alike than different.

In this chapter, you will appreciate the commonality of all life. We will review:

- Cell theory
- The three domains of life

- Containment—cell wall, cell membrane
- Cytoskeleton
- Cell organelles
  - Ribosomes
  - Endoplasmic reticulum
  - Mitochondria (prokaryotes living within eukaryotes?)
  - Chloroplasts
  - Golgi apparatus
  - Peroxisomes
  - Vacuoles and lysosomes
  - Proteasomes

Cell Theory

Needless to say, the cell theory is no longer a theory. We accept that all of life is composed of cells. (A debate lingers about those little anomalies called viruses that are simply packets of genetic material.) Perhaps you can image the amazement of the first human, when, not so long ago, he saw a living cell. Imagine Anton van Leeuwenhoek in the seventeenth century looking at pond water through a microscope and observing paramecium swimming in pond water. He must have felt like Neil Armstrong would have felt had he encountered something that moved on the lunar surface. Human nature being what it is, the observations of Leeuwenhoek and others took awhile to make an impact—about 200 years. By the middle of the nineteenth century, the cell theory was promulgated and accepted by the scientific community. According to cell theory, all organisms are made of cells and all cells come from preexisting cells.

The Three Domains of Life

There have been several reasonably successful attempts to define the divisions of life forms. At first, divisions of life forms were based on what we could see, and division into animals and plants seemed sufficient. However, the existence of unicellular organisms presented a problem. At first, they were classified as plants. Then only those with photoactive pigments were classified as plants. Then they were given a division of their own—protist. All of this was based on how things looked. Fungi were also tossed around divisions because they are unicellular, but not so primitive as bacteria.

Modern tools enable us to look at the fundamental structure of genetic material (in DNA and RNA) and derive relationships based on chemical similarities. The knowledge gained from similarities and differences in genetic material has changed the way we classify life forms. In 1970, Carl Woes of the University of Illinois examined the nucleotide sequence of the ribonucleic acid (RNA) contained in ribosomes of numerous organisms. The ribosome is a structure found in all living organisms and composes the factory that manufactures protein. It consists of proteins and a few molecules of RNA. The RNA is synthesized directly by the deoxyribonucleic acid (DNA) and is a direct reflection of the DNA code. Like DNA, RNA has only four different bases that comprise all RNA, but unlike DNA, RNA is single stranded. It's a nice molecule for comparative studies.

Dr. Woes found that all is not what it seems. Looking at the base sequences for ribosomal RNA and assuming that the more base sequences that two organisms have in common, the more closely related they are, he found that we have all derived from an ancient life form, no longer in existence. Bacteria appear to have diverged from everything else early in life's history. The world at the point of diversion consisted of bacteria and everything else. Then an odd group of unicellular organisms developed their own branch of life. These little critters look like bacteria, but chemically, they are closer relatives of multicellular organisms—that means us. The new group was called *archaea.*

In Archaea, the genetic transcriptions and translation—the way proteins are produced from the genetic code—are similar to eukaryotes and dissimilar to bacteria. However, Archaea's transfer RNA has features that differ from all living things. There are fundamental uniqueness in their cell membranes and cell walls.

Archaea range from 0.1 to 15 µm in diameter. Most are extremely small and only observable using an electron microscope. However, some form aggregates of filaments up to 200 µm in length.

Archaea were first thought to be remnants of the earliest inhabitants of the earth because they were initially found in extreme environments—hot springs, deep in the ocean, in petroleum deposits, highly saline places. However, recent discoveries of archaea among ocean plankton indicate that not only are they found in average environments, but also they may be the most abundant organisms on earth.

Dr Woes suggested that life could be divided into three *domains—bacteria,* archaea, and *eukaryotes.* The relationships between these domains can be described by a *phylogenetic tree* as shown in Fig. 2-1. According to this scheme, points where species diverge from one another are represented as branches in the tree.

Bacteria and archaea, although different domains, are both *prokaryotes,* because they lack a nucleus. The remaining domain consists of cells that have nuclei and are

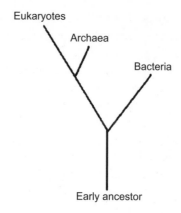

Figure 2-1 Phylogeny of life.

called *eukaryotes*. Eukaryotes include fungi, sponges, and us. The "karyotic" portion is derived from a Greek word meaning "kernel." The prokaryotic cells are so named because they preceded life forms wherein chromatic material is organized into a nucleus, that is, before cells had a kernel. Their nucleic acids are in the cytoplasm with everything else. Other cells are more advanced and are called eukaryotic because they have a nucleus. Eukaryotes also have numerous functional compartments within the cytoplasm called *organelles* which segregate the functions of the cell. The eukaryotic cells, therefore, are like a tidy office where papers are sorted onto desks and into file cabinets, as compared to the messy prokaryotes which appear like an office where piles of paper cover the floor. A depiction of eukaryotic versus prokaryotic cells is provided in Fig. 2-2.

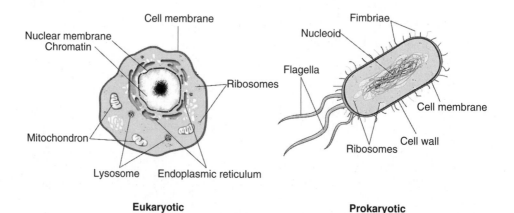

Figure 2-2 Eukaryotic versus prokaryotic cells.

Before we start feeling superior to our messy sister domains, note that prokaryotes have some advantages. They can live in diverse habitats and grow rapidly. Their rapid division rates allow them to evolve quickly and adapt to changing environments. Eukaryotes, though, can carry out more complex chemical reactions because they can isolate these reactions and place reactants closer together. Eukaryotic cells tend to be much bigger than prokaryotic cells because the prokaryotes depend on diffusion to obtain nutrients and eliminate much of the waste products. Eukaryotes can strategically locate import/export processes near the cell membrane and utilize transport mechanisms to move material around the cell. Archaea are smaller than bacteria. The smallest archaea are only about 200 nm across, compared to bacteria, which measure 1 to 10 μm. Eukaryotic cells can be huge—as large as 100 μm across.

Cell Containment

All cells have an oily, lipid *cell membrane*. Inside of this membrane exists a marvelous chemical system that allows each cell to grow, divide, fight off intruders, take in important reagents, and excrete other materials that are either waste products or important products of this particular cell's metabolism.

To understand the forces that create the membrane, envision a drop of oil in a container of water. The oil drops stay intact. Now add another drop of oil to the container. And then another. The droplets will coalesce, as if they are seeking each other's company and are avoiding the water. Chemicals that behave this way are called *hydrophobic*, or water-hating. The underlying reason for their behavior in water is that they have no electrical charge at their surface. As discussed in Chap. 1, water molecules are polar, that is, the molecule exhibits partial electrical charges. One end, the oxygen end, is electronegative. The other end (actually two ends), composed of the hydrogen molecules, is electropositive. The water molecules are attracted to each other, with the positive ends of some molecules seeking the negative ends of other molecules. As the partially charged ends try to get together, they force out anything that gets in the way, like an uncharged (oily) molecule, with the net effect that the oily molecules tend to clump together.

Molecules that bear charges behave differently in an aqueous solution. The charged areas (poles) of the other molecules compete with the water molecules for the opposite charges on the water molecules. As a result, the molecules migrate until they are evenly dispersed. From our vantage point, they disappear, or dissolve in the water. This type of molecule is *hydrophilic*, or water-loving. These molecules have polar ends.

The molecules that form cellular membranes are hybrids—one end is hydrophobic or oily (nonpolar) and the other end is hydrophilic, or water-loving (polar).

Molecules that exhibit different polarities are called *amphipathic*. The molecules line up in two layers because their nonpolar tails seek out one another, or, conceptually, are pushed together by water. The lipid tails cluster together with the tails of other molecules and induce the formation of a second layer. Polar heads protrude on either side of this double layer. Then, depending on their number and size, the molecular assembly curls around to form a ball. Any collection of amphipathetic molecules like this in an aqueous solution will behave the same way—line up tail to tail and with hydrophilic ends, projecting outside of the double-layered membrane. Cell membranes contain a variety of lipids with a high percentage of phospholipids. Cholesterol is also a major constituent of some cell membranes.

The ball is hollow, forming a type of micelle, called a *liposome*. The double-membrane of the liposome encloses a watery interior called the *cytosol*. This interior is isolated from the rest of the environment by its oily barrier. The chemical business of the cell can now be conducted in the privacy of the watery interior, safe from disruptive environmental forces. See Fig. 2-3.

Figure 2-3 The cell membrane and the cell as a liposome.

This is a great system because in order to dissolve in blood, a molecule must be water soluble, but in order to penetrate the cell, a molecule must be lipid soluble. The lipid barrier keeps out a lot of nasty stuff. The membrane allows some things to pass, such as small molecules of water or CO_2. In other words, the membrane is *selectively permeable.*

Most bioactive molecules need permission somehow to enter the cell. The way the permission is granted is through the existence of specific receptors that appear on the surface of the cell. The receptors are typically carbohydrates attached to a protein stem. The protein molecules that contain receptors float up through the membrane which behaves as a viscous semiliquid. The protein remains interior to the membrane but the receptor heads protrude above the surface. Each receptor is specific to a unique molecule. The receptors provide the entrance to the cell because the attached molecules can move through the membrane into the cell or be actively carried in. Sometimes the binding of a receptor to a polar molecule negates the charge. Once the molecule is nonpolar, the entire neutral molecule passes through the membrane readily. The type and amount of different receptors that are produced by the cell and that are transported to the surface are the primary mechanism by which the cell controls what and how much of what enters the cell. Consider insulin. Insulin regulates sugar metabolism within the cell. Insulin must attach to a receptor at the cell surface to be active. Therefore, the body can control sugar metabolism at two points—first, by controlling the amount of insulin that is produced and second, by controlling the amount that attaches to the cell membrane (proportionate to the insulin receptors that are on the surface of the cell).

Cells possess an internal scaffolding, called the *cytoskeleton*, that stabilizes the interior of the micelle. The cytoskeleton is composed of small protein fibers, called *microfibrils*. The load bearing microfibrils are made of strong proteins like keratin. Others form tubes of *microtubulin* that enable material transport. Still other microfibrils are contractile, providing a mechanism for cell movements. The latter are composed of actin or flagellin.

Most bacteria, archaea, algae, fungi, and plants have an additional protective layer called a *cell wall.* The cell wall is a rigid structure that functions to provide structural support, and protect the organism from environments with a different salt concentration than the cytoplasm. Without the cell wall, water would attempt to balance the salt concentration between the inside and the outside of the cell, and the cell would either shrink or burst. There is a great variety of cell wall compositions, dependent on the organism. In plants, the cell wall is made of a carbohydrate polymer, cellulose. Bacteria utilize a protein-carbohydrate complex called peptidoglycan. Fungi have walls of chitin and algae use glycoproteins and polysaccharides. Some algae even have walls composed of silicic acid. One of the distinguishing features of the archaea are their cell walls. The cell walls of archaea contain truly unique glycoproteins, pseudopeptidoglycan, or polysaccharides.

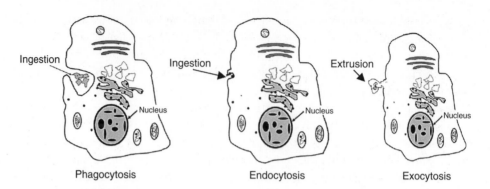

Figure 2-4 Eukaryotic cells can ingest and extrude material.

The cell wall is necessary to protect cells that do not enjoy the luxury of a surrounding that is *isotonic* (same salt concentration) with their innards. However, this wall means the cells cannot take in external material by *endocytosis*, a process by which a portion of the cell membrane infolds and pinches off with a bit of the external environment within it. The opposite process is unavailable also. Cells without walls can eliminate waste by a process called *exocytosis* wherein a vesicle containing unwanted material extrudes its contents to the exterior. Some eukaryotic cells can ingest a fair amount of extracellular material by surrounding it and engulfing it. Amoeboid cells called macrophages can eat entire bacteria through this process of *phagocytosis*. See Fig. 2-4.

Some bacteria have an extra membrane outside of their cell wall. This membrane was discovered because it prevented the cell wall from taking up a standard stain used to visualize the organism in the microscope. This stain is the *Gram stain*. Bacteria with the extra membrane do not take up the Gram stain and are called gram-negative. The common antibiotic, penicillin, attacks the exterior membrane of gram-negative bacteria. Penicillin is not effective against gram-positive bacteria.

Introduction to Cell Organelles

The interior of living cells is as compact as a submarine. Within the interior, bio-molecules whirl around rapidly in a random fashion. The achievement of chemical processes, whereby specific reagents must come together in an orderly fashion, is amazing indeed. This is especially true for prokaryotic cells, which have few options to isolate different processes. The eukaryotic cell is able to conduct certain bio-processes more efficiently than prokaryotes because different functions are packaged within interior structures, called *organelles*, that are rather like simplified cells themselves.

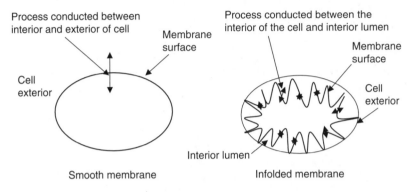

Figure 2-5   Advantages of infolded structures for process that occurs across
a membrane surface.

As we discuss characteristics of cells, you will be frequently confronted with
infolded, convoluted membranes. This common theme of cell structure is needed
because so many cell processes occur across a membrane surface. Such processes
include photosynthesis and oxidative metabolism. Therefore, cells with a greater
membrane surface area have an advantage. Further, the infolded structures
created interior chambers that concentrate needed enzymes and isolate processes.
See Fig. 2-5.

THE GENETIC MATERIAL

Eukaryotic cells have a double-layered inner membrane that protects the genetic
code in the DNA. This membrane is called the *nuclear membrane* and the entire
structure is known as the cell *nucleus*. The nucleus provides housing for the processes
involving DNA. Pores penetrate the nuclear membrane, enabling movement of
molecules to and from the nucleus. The interior of the nuclear membrane is connected
to thin fibrous sheets, called *lamina*, within the nucleus.

Eukaryotic cells have at least several chromosomes. Within the nucleus, each
chromosome is accorded a specific position where it is attached to the lamina. The
nucleus also contains a darkened area, called the *nucleolus*. The nucleolus is devoted
to the manufacture of the RNA found in ribosomes.

In a resting (nondividing) cell, the interior of the nucleus looks like a stringy mess.
The DNA is in the form of *chromatin*. As chromatin, the DNA forms coils around
protein balls called *histones*. There are light and dark regions within the chromatin.
The darker regions, called *heterochromatin*, contain DNA supercoils. Figure 2-6 shows
the development of supercoils. In the lighter regions, the *euchromatin*, the DNA is
less tightly coiled. Distinct chromosomes are not visible except during cell division.
See Fig. 2-6.

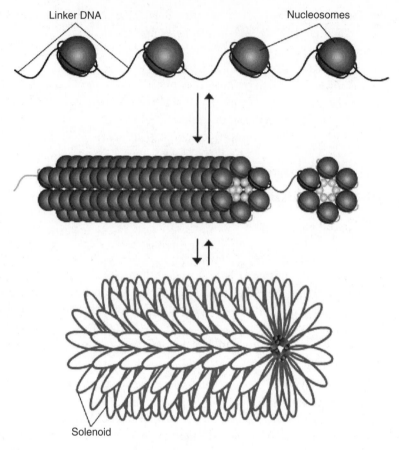

Figure 2-6   Chromatin.

Prokaryotic cells usually have only one chromosome. This chromosome is circular and is located in a darkened area called the *nucleoid*. There is no protection for this chromosome from the general milieu of the cytoplasm.

RIBOSOMES AND PROTEIN PRODUCTION

Ribosomes are found in all forms of life and are similar with all. Remember that the base sequences on the RNA within ribosomes formed the basis for the definition of the three domains of life—eukaryotes, bacteria, and archaea. Looking through a microscope, ribosomes look like pebbles distributed within the cytoplasm. Ribosomes are the factory units for the production of proteins.

In both types of cells, the ribosomes come in two subunits, one large and one small. The subunits do not unite unless the ribosome is actively producing proteins. Messenger RNA attaches to one of the ribosome subunits, bringing the DNA code. The other subunit then attaches and the ribosome interprets the code, producing protein. There are small differences between ribosomes of eukaryotes and prokaryotes, but both contain proteins and a few molecules of RNA.

ROLE OF THE ENDOPLASMIC RETICULUM

The *endoplasmic reticulum* (ER) is composed of a membrane in folds on the inside of the cell membrane. This membrane is continuous with the outer membrane of the nuclear membrane. The name is a mouthful. However, these structures were named when they were first observed and before the function was known. Early researchers observing this structure saw a folded structure (reticulum) protruding into the cytoplasm (endoplasmic). The folded configuration of the membrane provides a large surface area where chemical reactions can take place. As the membrane folds, it creates interior sacs.

In eukaryotic cells, ribosomes are frequently associated with the membranes of the ER. They make little knobs on the surface, giving the ER a rough appearance. You will hear this called the *rough endoplasmic reticulum*. Protein synthesis is intense on the rough ER surface. As proteins are being manufactured, they move to the interior (lumen) of the ER sacs where they undergo further processing. Among other things, proteins may require refolding to expose a hydrophobic surface. The hydrophobic surface allows them to float through the cell membrane. Proteins that are produced on the surface of the ER then migrate into the lumen and are extruded into the *Golgi apparatus* (see Fig. 2-9).

Some areas of the ER are smooth, lacking the knobby surface caused by ribosomes. The *smooth endoplasmic reticulum* is devoted to production, processing, and, in some cases, destruction of lipids.

Prokaryotic cells do not have organelles with huge surface areas provided by infolded membranes. However, they do compartmentalize the cytosol. The area immediately within the cell membrane is called the *periplasm* and concentrates some of the metabolic processes. Within the periplasm the inner surface of the cell membrane occasionally folds inward, forming convoluted structures called *mesosomes*. The mesosomes augment the surface area of the cell membrane to enable such necessary functions as the formation of a proton gradient for oxidative metabolism.

MITOCHONDRIA

You will learn that the energy of the cell is provided by an oxidation-reduction reaction, and the most efficient of this reaction utilizes free oxygen. Eukaryotic cells have a unique organelle, called a *mitochondrion*, in which the energy storage molecule (ATP) is produced using oxygen in a process known as oxidative phosphorylation. Mitochondria are the "powerhouses" of the cell. See Fig. 2-7.

Mitochondria are surrounded by a double membrane. In addition, mitochondria possess an inner membrane that is repeatedly folded inside the organelle. The folds are known as *cristae*. Like any surface folded inside a given space, the folds provide a large surface area, and it is upon across this surface area that the oxidative metabolism occurs. The enzymes necessary for the reactions of respiration are located within the cristae or within the matrix between the cristae. Prokaryotic cells, without the organelles known as mitochondria, are limited to the inside of the cell membrane to provide a substrate for oxidative metabolism.

Mitochondria are interesting because they appear to be an independent life-form. They generate the majority of the cell energy, including what they need for survival. Also, mitochondria contain their own DNA (called *mitochondrial DNA* or *mDNA*). Curiously, this mDNA is circular, like bacterial or archaeal DNA. Therefore, in addition to an energy source, mitochondria have their own information system. The mDNA codes for most, but not all, of the proteins needed by the mitochondria both for conducting oxidative metabolism and for maintaining the structure of the mitochondria. Furthermore, mitochondria duplicate themselves independent of cell

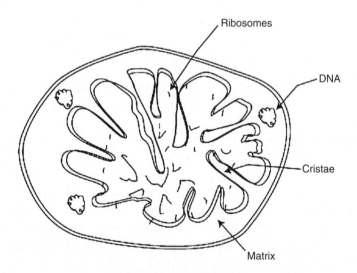

Figure 2-7   Mitochondria.

division. Perhaps they were once independent organisms, perhaps archaea, that invaded larger cells early in the evolution of life.

Mitochondrial DNA is of special interests in tracing relationships between groups of humans. We receive the progenitors of all of our cellular mitochondria from our mother. The maternal mitochondria are present in the egg when fertilization occurs. Sperm have no mitochondria. The paternal role in determining how our mitochondria function is limited to the few genes on the cellular DNA that are necessary for metabolism. The maternal mitochondria divide and proliferate and are distributed among the proliferating cells of the zygote. Mitochondrial DNA (mDNA) does not mix with the DNA from the paternal parent but remains independent of the nuclear DNA and also remains intact through the generations. We will discuss in Chap. 14 the usefulness of mDNA in determining relationships between groups of humans and in tracing *Homo sapiens* back to our origins.

Within the last 15 years or so, the cause of several rare, previously mysterious myoencephalopathies, or muscle wasting diseases, has been traced to mutant genes within the mDNA. Without fully functional mitochondria, victims of these diseases cannot produce enough energy. Such diseases are inherited strictly from the mother. Whether or not an individual progeny receives the mutated genes is luck of the draw. Usually only some of the maternal mitochondria contain the mutated gene; and the egg receives only a sampling of the maternal mitochondria. A child may have many, few, or none of the mutated mDNA. The inheritance of these diseases exhibits a random pattern, which explains why the cause remained mysterious for so long.

CHLOROPLASTS

Photosynthesis is conducted in eukaryotic cells within an organelle that is eerily similar to mitochondria. This organelle is the *chloroplast*. Like mitochondria, chloroplasts are surrounded by a double membrane. Unlike mitochondria, the inner folded structure is continuous with the inner membrane. The infolded membrane forms flattened vesicles called *thykaloids* stacked like pancakes into piles called *grana*. Interior to the thykaloids is the *stroma*. The thykaloids have embedded enzymes for performance of photosynthesis. The stroma also contains enzymes.

Like mitochondria, chloroplasts contain DNA. Like mitochondria, bacteria, and archaea, chloroplast DNA is circular. Another invasion?

Some bacteria have infoldings of the plasma membrane that support photosynthesis. These membranes contain the pigments and enzymes required and provide a surface for the establishment of a proton gradient that provides the energy to drive the process. See Fig. 2-8.

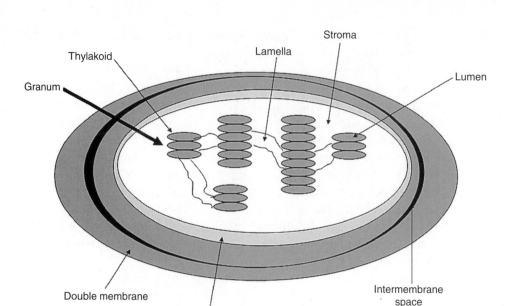

Figure 2-8   Chloroplasts.

GOLGI APPARATUS

The name, Golgi apparatus, has nothing to do with the function of the organelle. This structure was discovered by Dr. Camillo Golgi in 1898. The fact that it is a proper name also explains why we capitalize Golgi. Like the ER, the Golgi apparatus is composed of small internal compartments or sacs called *cisternae.* (Note that each different kind of sac, however similar they may be, has a different name.) The Golgi apparatus has a role in protein processing and in the; production, processing, and destruction of lipid; and in the production of lysosomes.

The Golgi apparatus receives proteins from the ER and extrudes them in the direction of the cell membrane. The organelle has a distinct orientation. The side that receives proteins is the *cis* side and the side that extrudes them is the *trans* side. Within the Golgi apparatus, proteins undergo further processing, such as refolding prior to extrusion through the cell membrane. The organelle receives and extrudes product in little vesicles that are extruded into the interior or meld with the membrane. The movement into and out of the apparatus makes the organelle surface appear to be bubbling. See Fig. 2-9.

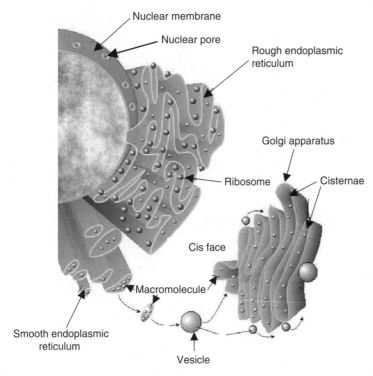

Figure 2-9   Movement of material between the endoplasmic reticulum, the Golgi apparatus, and the cytosol. (*Figure based on Wikipedia Commons*)

PEROXISOMES

I am sure you realize at this point that the eukaryotic cell is a mosaic of smaller, cell-like entities, some of which grow and divide on their own. One of these entities is the peroxisome. Like all cell organelles, peroxisomes are membrane-bound. Peroxisomes concentrate metabolic enzymes, specifically those enzymes that accomplish oxidative metabolism. They may also contain digestive enzymes. Each peroxisome is a specialist in that different individual organelles are devoted to different metabolic processes. For example, peroxisomes within the liver detoxify ethanol. Glyoxisomes, found in plants, contain enzymes that convert one of the products of photosynthesis into sugar. Seed peroxisomes contain enzymes that oxidize fatty acids to yield glucose and thereby provide an energy source.

Peroxisomes generate and detoxify free radicals. Their role in producing hydrogen radicals from hydrogen peroxide is important in general immunity against invading pathogens. Peroxisomes have another important role in immunity. They digest foreign proteins, preserving the foreign antigenic structure. In other words, they preserve the part of the foreign protein that the immune system recognizes. This antigenic structure is passed from the peroxisomes to immunologically active cells that activate an immune response. Drugs that target peroxisomes have had some preliminary success in treating autoimmune diseases. These drugs interrupt the immune response at the point where the body's proteins are first dissected into their antigenic structures. Normal immune systems do not respond to these self-antigens. Immune systems of individuals with autoimmune diseases respond to self-antigens.

VACUOLES AND LYSOSOMES

The issue of transporting material around the cell is somewhat problematic. Remember that the inside of the cell is friendly to water-soluble or hydrophilic molecules and not compatible with water-insoluble or hydrophobic molecules. However, inside the cell, molecules that are destined for the cell membrane may be hydrophobic, and, if not protected, would coalesce within the aqueous cytoplasm. The cell also needs protection from certain enzymes produced by the cell such as enzymes that are dedicated to waste disposal. These enzymes might digest essential cell structures. Eukaryotic cells have developed clever ways to transport molecules to their point of action. One method of moving material around is within small transport compartments surrounded by a hydrophobic (lipid) membrane. These compartments are called *vacuoles*. The vacuoles can transport material to and from the cell surface in a mode isolated from the rest of the cell interior.

Vacuoles can also be used for storage. Large percentages of plant cells can be occupied by vacuoles. Plant cell vacuoles may store energy in the form of starch. Also, plant cells use water stored in vacuoles to help offset changes in osmotic pressure in their environment. Water is released when the cytosol becomes too salty and is withdrawn into the vacuole when there is an excess of water.

There are many metabolic processes that produce excess organic molecules that the cell must eliminate. Such materials may be transported to the cell surface surrounded by a protective vacuole, which fuses with the cell membrane and extrudes its contents. Some vacuoles contain proteins from the cell surface, taken in by endocytosis, or also foreign material, ingested through phagocytosis. These and vacuoles containing internally generated waste molecules may fuse with a *lysosome* that contains digestive enzymes called *lysozymes*. Lysosomes are formed when pieces of the Golgi apparatus are pinched off as a sac containing the lysozymes. See Fig. 2-10.

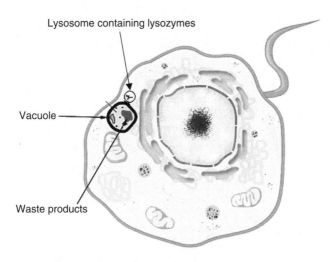

Figure 2-10   Lysozymes are dumped into a vacuole to digest waste molecules.

PROTEASOMES

The treatment of degraded or improperly processed proteins presents a special challenge. Many of the building blocks for proteins, amino acids, are essential dietary components. Others are energy-intensive in their construction. The ability to recycle amino acids is important. Also, proteins that are not properly configured can cause a great deal of mischief within the cell and must be quickly eliminated before they precipitate or otherwise interfere with cellular processes. *Proteasomes* are organelles devoted to the dismantling of degraded or misfolded proteins. The proteasome recognizes a tag, called *ubiquitin*. Ubiquitin attaches to surface markers that indicate the protein is improperly configured. Proteasomes are constructed of α- and β-subunits. The exterior of the proteasomes is composed of the α-subunits, which have a receptor for the ubiquitin. Once attached to the proteasome, the marked protein is drawn into the interior of its barrel-like structure, composed of β-subunits, which house proteolytic enzymes. The protein is dismantled and the amino acids are released for reuse by the cell. See Fig. 2-11.

Summary

Cells come in two types. Eukaryotic cells contain their genetic material within a nucleus surrounded by a nuclear membrane. They are filled with subcompartments within their interior, called organelles that concentrate specific metabolic functions. Prokaryotic

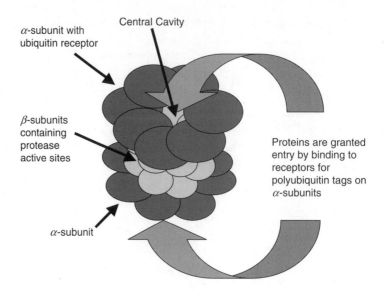

Figure 2-11 Proteosome.

cells have no nucleus and no interior organelles. Because they are less efficient and have poorly developed means for transporting material into and out of the cell, prokaryotic cells are as a rule, much smaller than eukaryotic cells. Prokaryotic cells consist of bacteria, and a distinctive group of organisms called archaea. In spite of the physical similarities, archaea are phylogenetically closer relatives of eukaryotes than of bacteria.

Both types of cells are surrounded by a double-layer cell membrane. The membrane molecules have carbohydrate heads and lipid tails. The hydrophobic molecules within the membrane keep environmental water out and cellular water in. Literally floating in the membrane are proteins with carbohydrate heads. The proteins float up to the outer surface and with the carbohydrate heads protruding out into the environment. These heads comprise receptors that are specific to a given target molecule. Once the molecule is bound to the receptor, it can be taken into the cell.

Proteins, algae, and plants are further contained within a rigid cell wall. The cell wall of plants is composed of cellulose. Other cell walls contain various glyco-proteins. The cell wall protects the organism from a potentially hostile and certainly nonisotonic environment.

The eukaryotic cell organelles are devoted to energy production (mitochondria and chloroplasts), biomolecule production (rough and smooth ER, Golgi apparatus, peroxisome), biomolecule modification (rough and smooth ER and Golgi apparatus), digestion/destruction (peroxisomes, lysosomes, and proteasomes) and storage (vacuoles). Mitochondria and chloroplasts both contain a small amount of DNA upon a circular chromosome. Both divide independently of the cell. Mitochondrial DNA is of interest in tracing relationships between groups because it is passed directly from mother to child in the cytoplasm of the egg. Mitochondria and chloroplasts contain internally folded membranes and interior sacs that enable oxidative respiration and photosynthesis, respectively. The endoplasmic reticulum also consists of folded membranes, continuous with the nuclear membrane. The rough endoplasmic reticulum exhibits ribosomes on its surface and is important in protein production and processing. The smooth endoplasmic reticulum is important in lipid production and processing. The Golgi apparatus receives material from the endoplasmic reticulum and engages in further processing especially of proteins. The Golgi apparatus also produces lysosomes and the digestive enzymes, lysozymes, within lysosomes. Lysosomes can dump their enzymes into vacuoles that contain waste material, accomplishing digestion of this material.

Among the organelles that divide independently are peroxisomes. Peroxisomes concentrate metabolic processes that require oxygen. Different peroxisomes perform different functions, specializing in specific metabolic reactions. Proteasomes are devoted to the digestion of waste proteins. They allow recycling of the amino acid components of the proteins.

Quiz

1. Archaea were determined to be a unique life-form by
    (a) the fact that they live in extreme environments.
    (b) the molecules in their cell wall.
    (c) their small size.
    (d) the unique sequences on their ribosomal RNA.
    (e) All of the above.

2. Cell walls are found
    (a) only in bacteria.
    (b) only in gram-positive bacteria.
    (c) only in prokaryotic cells.
    (d) only in organisms that need to regulate influx of water from their environment.
    (e) c and d.

3. The cell membrane helps to regulate the influx and efflux of material
    (a) by actively excluding nonpolar molecules.
    (b) by attracting nonpolar molecules to the nonpolar molecules on the membrane surface.

(c) by selective permeability.

   (d) by allowing free diffusion.

4. The DNA within mitochondria

   (a) codes for all the proteins needed for oxidative metabolism.

   (b) is pinched off the major chromosomes.

   (c) is carried from the mother via the egg.

   (d) Both a and c are correct.

   (e) None of the above.

5. Lysosomes

   (a) destroy proteins.

   (b) contain digestive enzymes.

   (c) generate hydrogen peroxide.

   (d) All of the above.

6. Cell movement is provided by

   (a) components of the cytoskeleton.

   (b) contractile fibers within the cell membrane.

   (c) changes in osmotic pressure in different parts of the cell.

   (d) All of the above.

7. Prokaryotic cells

   (a) have no order within their cytosol.

   (b) do not have ribosomes.

   (c) do not have mitochondria.

   (d) do not have infolded structures.

   (e) All of the above.

8. The following is true regarding animal and plant cells:

   (a) Only animal cells have mitochondria.

   (b) Both contain vacuoles.

   (c) Plant cells have cell walls.

   (d) Both b and c are correct.

   (e) All of the above.

9. The Golgi apparatus

   (a) receives proteins from the rough ER.

   (b) produces lysosomes.

   (c) processes lipids.

   (d) conducts oxidative metabolism.

   (e) a, b, and c are correct.

   (f) All of the above.

10. Digestive enzymes are isolated to prevent damage to cell structures by

   (a) embedding the enzymes on infolded membranes.

   (b) isolating the enzymes within lysozymes.

   (c) isolating the enzymes and substrates within vacuoles.

   (d) conducting digestion exterior to the cell.

   (e) All of the above.

   (f) b, c, and d are correct.

   (g) Both b and c are correct.

CHAPTER 3

Carbohydrates

Carbohydrates, molecules consisting of carbon, hydrogen, and oxygen, are the most abundant biological molecule on earth, playing a role in such diverse functions as energy storage (glucose) and structure formation (cellulose and chitin). In addition, they can be found as components of the nucleic acids, and derivatives of carbohydrates can be found in certain coenzymes. We have listed just a few of the important roles that these long-chain carbon molecules play in the game of life. In fact, the importance of carbohydrates in biological systems cannot be overstated.

Carbohydrates are also called *saccharides* and all carbohydrates can be built up from a simple carbohydrate called a *monosaccharide*. If a molecule contains two monosaccharides, we say it is a *disaccharide*. *Polysaccharides* contain multiple monosaccharides (more than two). An *oligosaccharide* is a polymer consisting of between 2 and 20 monosaccharides, so typically the term polysaccharide is reserved for a molecule containing more than 20 monosaccharides. Colloquially, a saccharide is called a *sugar*.

If a derivative of a carbohydrate is attached to another molecule such as a protein or lipid, we say it is a *glycoconjugate*. Let's begin our discussion with a description of the simplest carbohydrate building blocks.

Monosaccharide

Like other biomolecules, carbohydrates exhibit a pattern of modular assembly. In this case, the fundamental unit is a monosaccharide which is a molecule with between 3 and 8 carbon atoms together with a functional aldehyde or ketone group. At room temperature, these are crystalline solids that are white in color. We can learn about the structure of carbohydrates in general by learning how monosaccharides are put together. A carbohydrate contains carbon (C), hydrogen (H), and oxygen (O). The basic building block of all carbohydrates is called a monosaccharide which can be described by the simple formula

$$C_nH_{2n}O_n$$

This formula is sometimes written more compactly as $\left(CH_2O\right)_n$. In either case, we see that *a key characteristic of a monosaccharide is the number of carbon atoms it contains*. The most important monosaccharides of biological importance are those with $n = 3, \ldots, 6$, that is, molecules that contain between 3 and 6 carbon atoms. These are named using a prefix that indicates the number of carbon atoms and ending with "*-ose.*" Hence *triose* is a monosaccharide containing 3 carbon atoms which can be described by the formula

$$C_3H_6O_3$$

In Table 3-1, we include a listing of the simplest monosaccharides and their formulas.

Table 3-1   Names and Formulas of Monosaccharides

Monosaccharide	Formula
Triose	$C_3H_6O_3$
Tetrose	$C_4H_8O_4$
Pentose	$C_5H_{10}O_5$
Hexose	$C_6H_{12}O_6$
Heptose	$C_7H_{14}O_7$
Octose	$C_8H_{16}O_8$

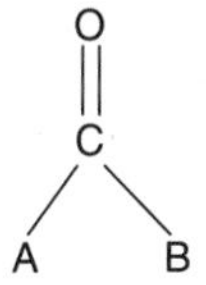

Figure 3-1 A carbonyl group.

Monosaccharides can also be further classified by the placement of its *carbonyl group*. Recall that a carbonyl group consists of a carbon atom which is connected to an oxygen atom with a double bond. This is illustrated in Fig. 3-1 where A and B can be other atoms. Note that all molecules depicted in this chapter are *Fischer projections*, a method of drawing molecules developed by Emil Fischer who studied carbohydrates in the late nineteenth century. A Fischer projection is essentially a two-dimensional projection of a three-dimensional molecule onto a flat sheet of paper (see Chap. 1).

If the monosaccharide is called an *aldose*, then the carbonyl group is an *aldehyde*. This is simply a carbonyl group where one of the attached atoms is a hydrogen atom, as illustrated in Fig. 3-2.

Classification of a molecule as an aldose can be combined with the number of carbon atoms to completely characterize the molecule. For example, *aldopentose* is a monosaccharide with 5 carbon atoms, one of which forms an aldehyde group. A very important aldopentose is called *ribose*, which forms a fundamental component in ribonucleic acid (RNA). For example, D-ribose is shown in Fig. 3-3.

The carbonyl group in a monosaccharide can also be a *ketone*. You can recognize this by the presence of a ketone group on the carbohydrate that takes the following form.

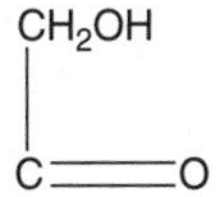

When drawing a Fischer projection, we place the aldehyde or ketone group at the top and proceed downward vertically along the carbon atoms that make up the chain of the molecule. The carbon atoms are numbered, starting with carbon number 1 at the top. This is illustrated in Fig. 3-4.

Figure 3-2 An aldehyde.

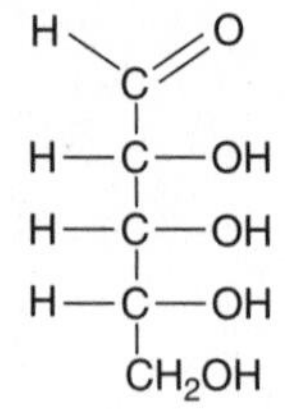

$$\begin{array}{c} H\diagdown\ \diagup O \\ C \\ | \\ H-C-OH \\ | \\ H-C-OH \\ | \\ H-C-OH \\ | \\ CH_2OH \end{array}$$

Figure 3-3 An example of an aldopentose, D-ribose, a component of ribonucleic acid (RNA).

D- AND L-MONOSACCHARIDES

We recall from our earlier studies in chemistry that *isomers* are molecules that have the same chemical formula and the same bonds, but the configuration of the atoms is different. The configuration of atoms in a molecule can vary in two basic ways. For example, we could have two molecules with the same chemical formula but can rearrange the order of the atoms. Another possibility is by orienting the atoms differently in space, but keeping the same order in both molecules.

When the atoms of a given isomer are kept in the same order but their arrangement in space is different, we call this a *stereoisomer*. If a stereoisomer is called an *enantiomer*, this means that two stereoisomers are mirror images of each other. Enantiomers are classified as D or L according to their *chirality*. Chiral molecules interact with plane polarized light, with D and L isomers rotating the light in opposite directions. The details of this are not important to us, but note that D isomers and L isomers react differently in the body.

A *chiral carbon* is one that has four different R groups attached to it. We identify an isomer as being in the D or L configuration by examining the chiral carbons in the compounds Fischer projection. We can see how this designation works with a simple hypothetical example. Imagine a carbon atom connected to a group R, a group R′, a hydrogen atom, and some other constituent that we designate by X. This

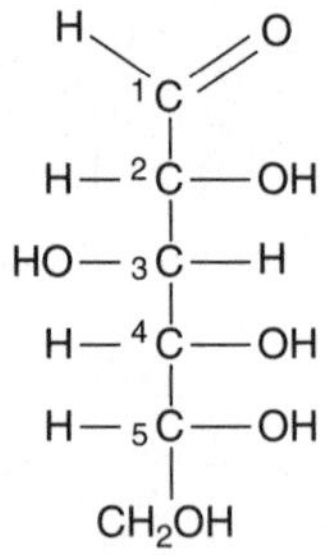

$$\begin{array}{c} H\diagdown\ \diagup O \\ {}_1C \\ | \\ H-{}_2C-OH \\ | \\ HO-{}_3C-H \\ | \\ H-{}_4C-OH \\ | \\ H-{}_5C-OH \\ | \\ CH_2OH \end{array}$$

Figure 3-4 In a Fischer projection, carbon atoms are numbered from the top. These numbers are not usually shown explicitly, but are understood.

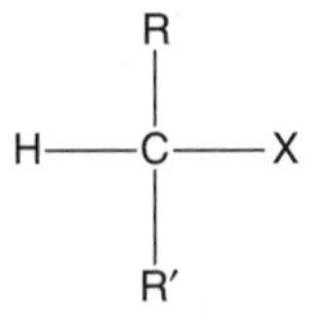

$$\begin{array}{c} R \\ | \\ H\!\!-\!\!-\!\!C\!\!-\!\!-\!\!X \\ | \\ R' \end{array}$$

Figure 3-5　A D-isomer configuration.

is a chiral carbon because it is connected to four different R groups. The D isomer has the hydrogen atom to the left, and X to the right. This is depicted in Fig. 3-5.

To obtain the L-isomer configuration of the molecule, we simply rotate or flip X-H about the carbon atom. The L isomer corresponding to the D isomer in Fig. 3-5 is depicted in Fig. 3-6.

This concept is easily extended to complex molecules, which can have multiple chiral carbon atoms. If a molecule has n chiral carbons, then it can have up to 2^n stereoisomers. A D-monosaccharide is one such that the OH group of the second to last or *penultimate* carbon atom is shown on the right-hand side (e.g., if a monosaccharide contains 5 carbon atoms, we check the position of the OH group on carbon number 4). Hence the OH group in a monosaccharide plays the role that X does in Figs. 3-4 and 3-5.

In Fig. 3-7, we show a 6-carbon monosaccharide in the D configuration. Can you guess what the name of this molecule is? Since it has 6 carbon atoms, it is a *hexose*. The carboxyl group at the end is an aldehyde, making this an aldose. Combining these two designations we can refer to this molecule as an *aldohexose*. In fact this is a molecule we all know and love, *glucose*.

It is easy to see why this is a D isomer. Looking at the fifth carbon atom counting down from the carboxyl group at the top, we see that the OH group is to the right of this carbon atom. Therefore we designate this molecule as *D-glucose*. This molecule is sometimes called *dextrose* or *blood sugar* and it is the most abundant organic compound on earth. D-glucose is the fundamental monosaccharide unit used in the construction of starch and cellulose.

Taking a closer look at D-glucose, notice that the molecule contains multiple chiral carbon atoms. The top carbon in the carboxyl group is *not* a chiral carbon because it is not attached to four different R groups. However, carbons 2 to 5 are chiral. In Fig. 3-8, on the left side we have circled the R groups for carbon number 2, while on the right we have circled the R groups for carbon number 4.

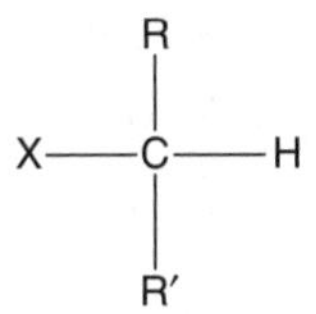

$$\begin{array}{c} R \\ | \\ X\!\!-\!\!-\!\!C\!\!-\!\!-\!\!H \\ | \\ R' \end{array}$$

Figure 3-6　An L-isomer configuration.

Biochemistry Demystified

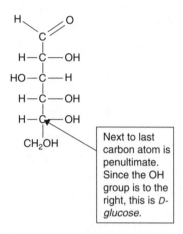

Figure 3-7 An illustration of D-glucose.

These molecules are a bit more complicated than the schematic example used to generate Figs. 3-4 and 3-5, but we use the same procedure when converting a D configuration to an L configuration. In a molecule with a long chain of carbon atoms, we flip the R groups to the left and right of each chiral carbon. So, to obtain L-glucose, we take the mirror image of the molecule in Fig. 3-7, *leaving out the top and bottom carbon atoms*. When this is done in the case of glucose, the OH group on the penultimate carbon atom appears on the left-hand side. This is called L-glucose, which is illustrated in Fig. 3-9.

While L-isomers are possible, most carbohydrates seen in nature are D isomers. Enantiomers show marked differences when the molecules are involved in reactions with other chiral molecules or chiral enzymes. If a reaction takes place with the assistance of a chiral enzyme, it is usually the case that one enantiomer will react rapidly while the other will not. In fact, in many cases the other enantiomer will not react at all. This is because the three-dimensional shapes of the two enantiomers are

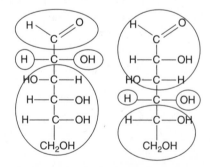

Figure 3-8 A chiral carbon has four different R groups attached to it.

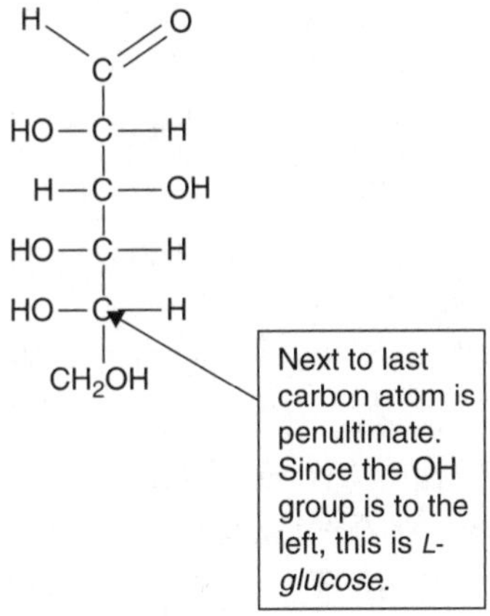

Next to last carbon atom is penultimate. Since the OH group is to the left, this is *L-glucose.*

Figure 3-9   An example of an L-isomer, L-glucose.

different, and a chiral enzyme can only match up with one of the configurations. We can think of the differences between enantiomers and how they work with chiral enzymes using a lock and key analogy. The enantiomer plays the role of a key while the chiral enzyme is the lock. The D-isomer and L-isomer are different keys, and typically only the D-isomer fits the lock (the chiral enzyme) and can be utilized. A good example of this involves the metabolism of glucose, which has enantiomers D-glucose and L-glucose. While the body utilizes D-glucose for fuel, it cannot metabolize L-glucose at all. However, not all molecules important in biochemistry are D-isomers. For example, a compound called L-dopa can be used to treat Parkinson's disease. However, the D-dopa isomer has no effect.

A *diastereomer* is a stereoisomer that is not a mirror image. The best way to illustrate this is by example. The compound 2,3-dichloropentane has two enantiomers with the chlorine atoms on the same side, as shown in Fig. 3-10.

However, 2,3-dichloropentane also has two more stereoisomers, with each chlorine atom placed on the opposite side of each chiral carbon. These are shown in Fig. 3-11. Notice that neither molecule from Fig. 3-11 can be superimposed on the configurations shown in Fig. 3-10.

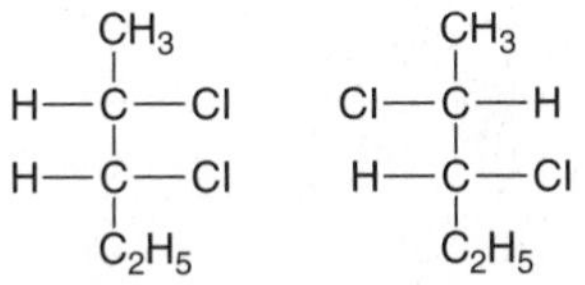

Figure 3-10   Two of the stereoisomers of 2,3-dichloropentane.

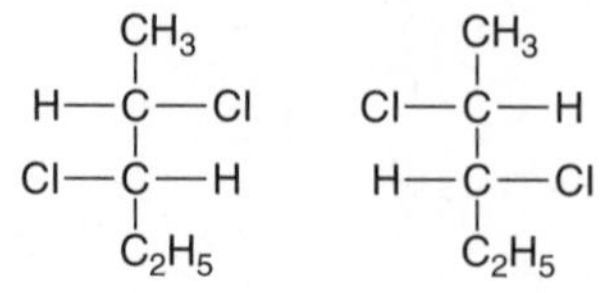

$$CH_3 \qquad CH_3$$
$$H-C-Cl \qquad Cl-C-H$$
$$Cl-C-H \qquad H-C-Cl$$
$$C_2H_5 \qquad C_2H_5$$

Figure 3-11 Two more stereoisomers of 2,3-dichloropentane. These stereoisomers are diastereomers of the compounds shown in Fig. 3-10. Neither molecule from Fig. 3-11 can be superimposed on the molecules in Fig. 3-10 or vice versa.

Diastereomers also occur in carbohydrates. A common examples is the diastereomer pair of D-ribose and D-xylose. We have already seen D-ribose in Fig. 3-3. To obtain L-ribose, we would swap the OH groups and H atoms on every chiral carbon. This is shown in Fig. 3-12.

Now, consider generating a different stereoisomer from D-ribose while leaving the hydroxyl group on the penultimate carbon atom in the same position (to the right of the carbon atom). This would generate another D-isomer, but one with a different structure. If we flip the hydroxyl group and the hydrogen atom on carbon number 3, we obtain D-xylose, which is a diastereomer of D-ribose. This is illustrated in Fig. 3-13.

An *epimer* is a molecule whose configuration differs from another molecule only at a single asymmetric carbon atom called the *epimeric carbon*. D-Glucose and D-mannose are epimers. In glucose, the OH group on the number 2 carbon is on the left of the third carbon atom, but in mannose it is on the right.

CYCLIC STRUCTURES OF MONOSACCHARIDES

While it is useful to learn about the structure and forms of monosaccharide using the straight-chain Fischer diagrams that we have used up to this point, this is not quite the whole story. Monosaccharides contain aldehydes and ketones together with a carbonyl group at the other end. These components can and do react

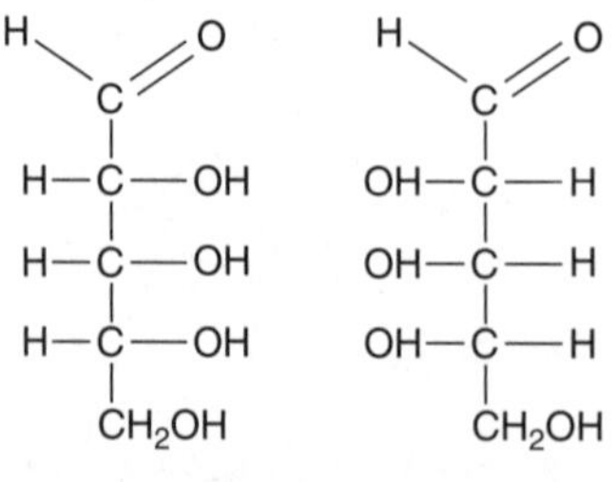

Figure 3-12 D-Ribose, on the left, compared with L-ribose.

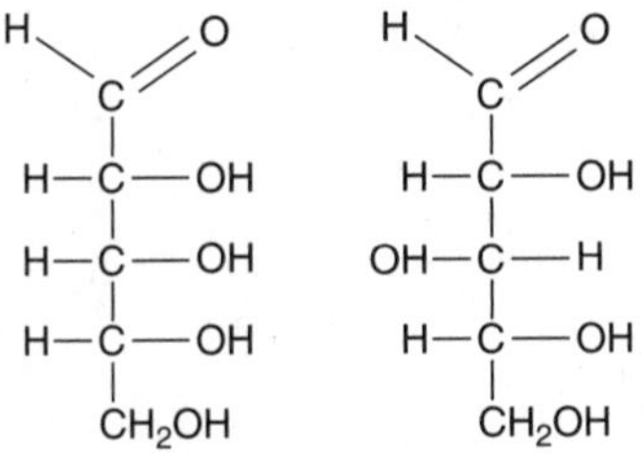

Figure 3-13   A comparison of two diastereomers. D-ribose is on the left, and D-xylose is on the right.

intermolecularly to form cyclic compounds. A carbohydrate with a carbonyl group (i.e., an aldehyde) can form a *cyclic hemiacetal* by bonding with a hydroxyl group elsewhere on the molecule. If a carbohydrate with a ketone group undergoes an intermolecular reaction with the carbonyl group, the result is a *cyclic hemiketal*.

A *pyranose ring* is a six-membered ring-structured monosaccharide. Such a ring can be formed from D-glucose by the addition of the OH group at carbon-5 to the carbonyl group at carbon 1. This is illustrated in Fig. 3-14.

The result of this process is a six-membered ring formation that can exist in two isomers because carbon-1 becomes chiral. There are two possibilities. Approximately 36% of glucopyranose rings in an aqueous solution are in α-configuration, where the OH group is shown below carbon-1, as illustrated in Fig. 3-15.

Approximately, 64% of glucose rings are in the β-configuration in aqueous solution, where we find the OH group above the first carbon. This is illustrated in Fig. 3-16.

The representations of the cyclic-carbohydrate structures as shown in Figs. 3-15 and 3-16 are called *Haworth projections*.

The isomer α-D-glucose and the isomer β-D-glucose are diastereomers. However note that they have the same configuration at all chiral carbons except at carbon-1s, where the positions of the H and OH are reversed. Pairs of molecules related this way are called *anomers*.

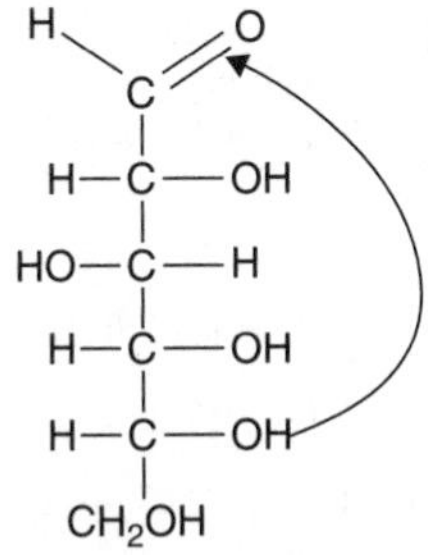

Figure 3-14   To form a pyranose ring from D-glucose, the hydroxyl group on carbon-5 is added to the carboxyl group at carbon-1.

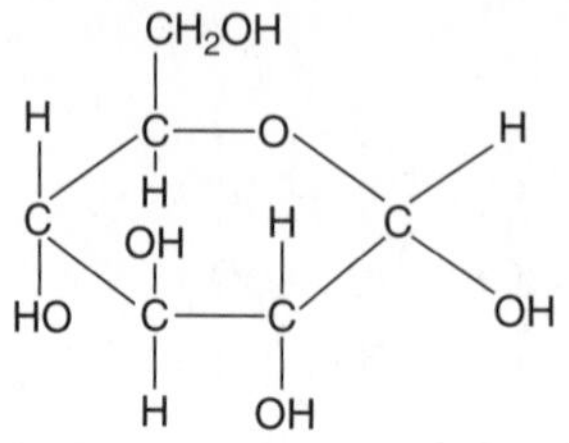

Figure 3-15 The isomer α-D-glucose.

A five-membered carbon ring is called a *furanose*. As an example, we consider D-ribose. The OH group on carbon-4 will react with the carbonyl group of carbon-1. This is illustrated in Fig. 3-17.

The same α, β designation is used depending on where the OH group is positioned relative to carbon-1 in the resulting ring. In Figs. 3-18 and 3-19, we see the Haworth projection of α-ribo-furanose and β-ribo-furanose, respectively.

After the formation of the ring, the first carbon is called the *hemiacetal carbon* if the open-chain compound that formed the ring was an aldose. If the open-chain form of the molecule was a ketose, then the first carbon in the ring form is called the *hemiketal carbon*. Sometimes the carbon-1 is called the *anomeric carbon*.

The reactions that form these carbon rings are reversible and occur spontaneously. The β-configuration is preferred because the large OH and CH_2OH groups lie in the plane of the six-membered ring. In α-configuration, the OH group that is shown below the anomeric carbon is perpendicular to the ring and feels repulsive forces from the H atoms in its vicinity. This is a higher energy state and the molecule would rather be in the lower energy β-configuration.

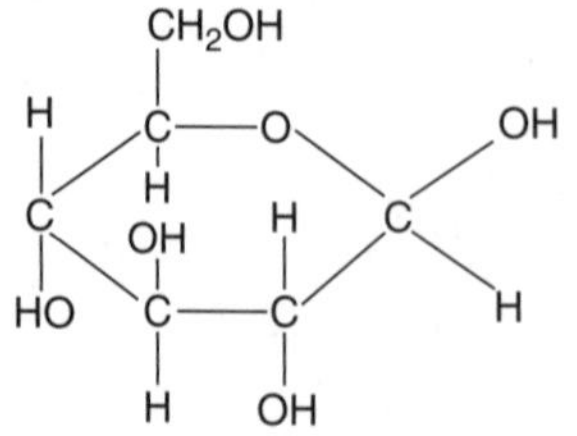

Figure 3-16 The isomer β-D-glucose.

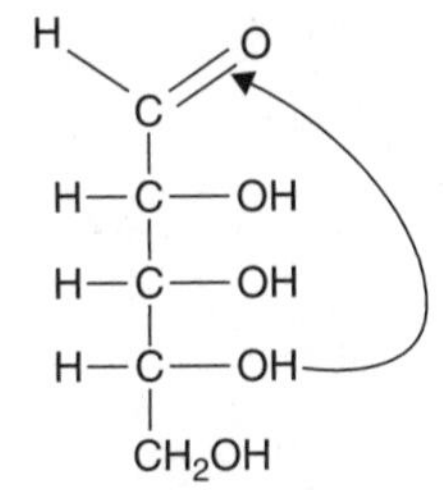

$$
\begin{array}{c}
\text{H} \diagdown \\
\text{C} = \text{O} \\
\text{H} - \text{C} - \text{OH} \\
\text{H} - \text{C} - \text{OH} \\
\text{H} - \text{C} - \text{OH} \\
\text{CH}_2\text{OH}
\end{array}
$$

Figure 3-17 To form a furanose ring from D-ribose, we react the OH group on carbon-4 with the carboxyl group.

Reducing and Nonreducing Sugars

If a carbohydrate has a free anomeric carbon that can be oxidized or undergo other reactions, the molecule is a *reducing sugar*. These compounds can reduce $Cu^{2+} \rightarrow Cu^{+}$. Examples include glucose and lactose. In contrast, consider sucrose, which has both anomeric carbons in a glycosidic bond (see below) and so it is nonreducing.

Derivatives of Carbohydrates

Several derivative molecules of carbohydrates exist and can be formed by reduction. For example, if a given molecule has one fewer oxygen then we say that it is a *deoxycarbohydrate*. The most famous of these molecules is *deoxyribose*, which is the primary constituent of deoxyribonucleic acid (DNA). This sugar differs from D-ribose in that one of the hydroxyl groups is replaced by a hydrogen atom (removing one oxygen) as shown in Fig. 3-20.

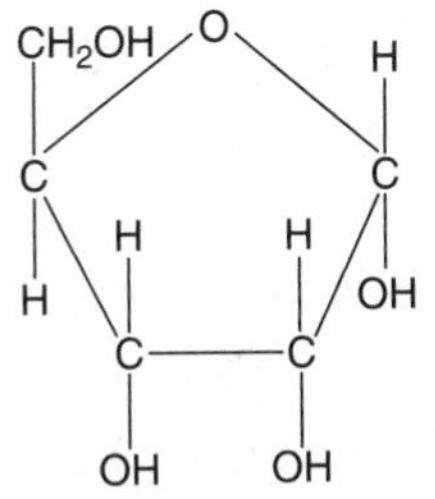

Figure 3-18 The Haworth projection of α-ribo-furanose.

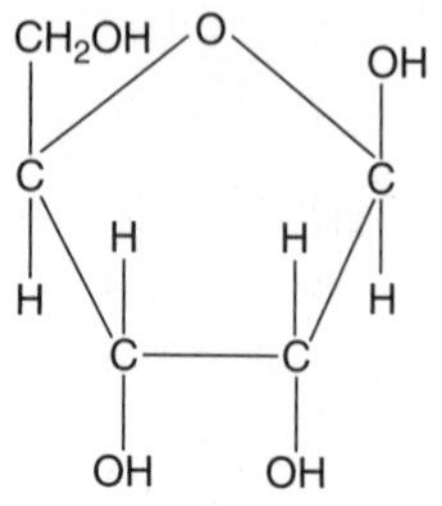

Figure 3-19 The Haworth projection of β-ribo-furanose.

A *sugar alcohol* is formed when the carbonyl group is reduced to a hydroxyl group. Sugar alcohols can be useful as a dietary substitute for a lot of reasons. They have less calorie content than ordinary sugar, and cannot be digested by bacteria present in the mouth. As a result candy that is sweetened with a sugar alcohol is less likely to contribute to tooth decay. For this reason many products such as chewing gum are sweetened with sugar alcohols. An important use of sugar alcohols is in the manufacture of diabetic-friendly foods. Sugar alcohols are only partially absorbed in the small intestine resulting in smaller changes in blood-glucose levels. Moreover, sugar alcohols that are absorbed are partially excreted in the urine rather than being fully metabolized.

Sorbitol is formed from glucose by changing the aldehyde group into a hydroxyl group. Sorbitol is illustrated in Fig. 3-21.

Other derivatives of carbohydrates are possible. A sugar amine is a molecule obtained from a carbohydrate by substitution of an amine group for a hydroxyl group. For example, *glucosamine*, which is a major component in the structure of crustacean shells and a precursor of glycosaminoglycans found in the joints of mammals, has the chemical formula

$$C_6H_{13}NO_5$$

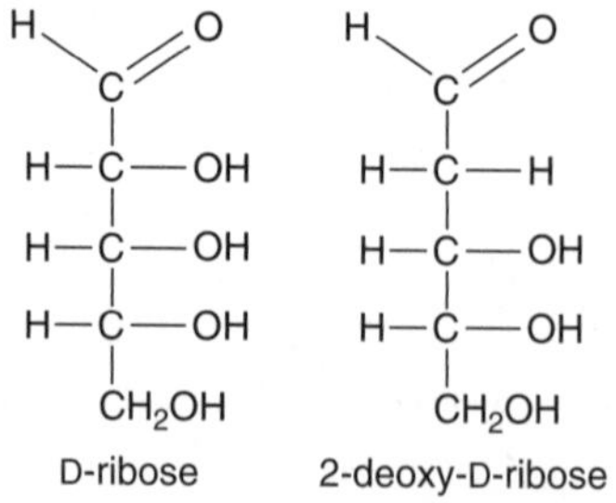

D-ribose   2-deoxy-D-ribose

Figure 3-20 A comparison of D-ribose and 2-deoxy-D-ribose. The deoxycarbohydrate is obtained by replacing the hydroxyl group on the second carbon atom with a hydrogen atom.

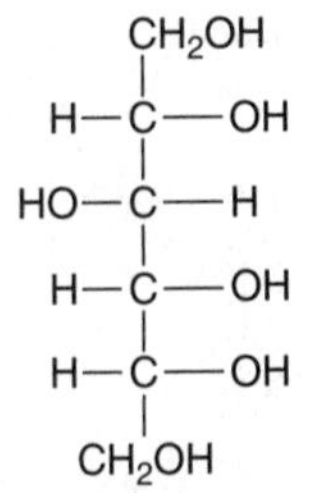

$$CH_2OH$$
$$H-C-OH$$
$$HO-C-H$$
$$H-C-OH$$
$$H-C-OH$$
$$CH_2OH$$

Figure 3-21   An example of a sugar alcohol called *sorbitol*. Compare with glucose, shown in Fig. 3-7.

It is obtained from glucose with the substitution of an amine group, NH_2 for a hydroxyl group, OH. Glucosamine is shown as a Haworth projection in Fig. 3-22, and in a three-dimensional rendering in Fig. 3-23.

A *sugar acid* can be formed when the first carbon or the carbon in the carboxyl group of a monosaccharide is oxidized. When the first carbon of glucose is oxidized, the result is a compound called *gluconic acid.* Since we get the acid by oxidation and the chemical formula for glucose is $C_6H_{12}O_6$, the chemical formula for gluconic acid is $C_6H_{12}O_7$. Gluconic acid is shown in Fig. 3-24.

Another important carbohydrate derivative is ascorbic acid. The L-enantiomer of this compound is vitamin C, which cannot be produced by primates and so you, biochemist, must obtain it from the diet. Ascorbic acid is a derivative of gluconic acid which we met in the last paragraph.

Glycosidic Bond and Disaccharides

When two molecules join together to form a new, single molecule while giving up or releasing a small molecule, we say that a *condensation reaction* has occurred. Condensation reactions are important in the chemistry of carbohydrates, leading to

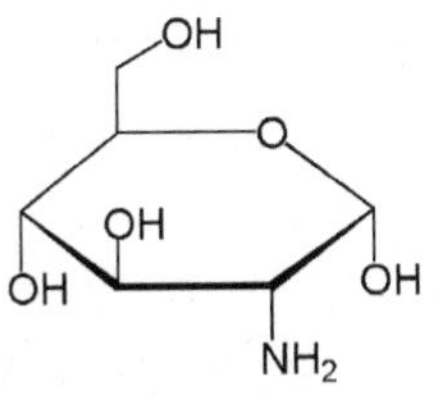

Figure 3-22   Glucosamine, which is obtained from glucose by replacing a hydroxyl group with NH_2.

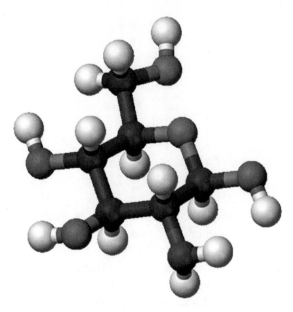

Figure 3-23 A three-dimensional rendering of glucosamine.

the formation of more complex molecules called *disaccharides* that are two monosaccharides bonded together.

One way that carbohydrates can react is via the formation of a *glycosidic bond*. A *glycoside* is a carbohydrate for which the OH group attached to the anomeric carbon has been replaced by an –OR group. For example, α-D-glucose can react with methanol to produce a compound called methyl α-D-glucoside. This is a condensation reaction that takes place in the presence of an acid catalyst, with water being given off as a by-product. The end result is that the OH group attached to the anomeric carbon is replaced by OCH$_3$. This is shown in Fig. 3-25. Note that the reaction liberates a water molecule, as shown in the figure.

Glucose

Gluconic acid

Figure 3-24 Gluconic acid is formed by the oxidation of the first carbon in glucose.

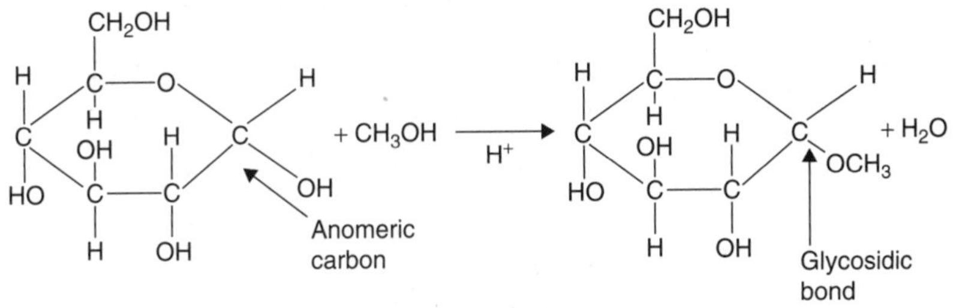

Figure 3-25   The reaction of glucose with methanol in the presence of an acid yields methyl α-D-glucoside.

In this type of reaction, the anomeric carbon and its two –OR groups constitute the glycosidic bond. Next, we consider a similar reaction in which glucose + ethanol yield a compound called *ethyl glucoside* plus water. This is shown in Fig. 3-26.

Glycosides obtained from 6-carbon rings like glucose are called *pyranosides*. When the monosaccharide participating in the reaction is a furanose, we say that the resulting glycoside is a *furanoside*.

Glycosidic bonds are important in building up more complex sugars from monosaccharides. In this context, a glycosidic bond is analogous to peptide bonds which join amino acids together to form proteins. The simplest molecule that can be constructed in this way is a disaccharide, or 2-sugar molecule in which one of the monosaccharides is linked via the OH group connected to the anomeric carbon. A familiar disaccharide is ordinary table sugar, which is called *sucrose*. This compound is a disaccharide molecule formed from glucose and fructose and can be obtained in nature from sugar cane and beets. The linkage is between the C-1 or anomeric carbon of α-D-glucose and the second carbon (C-2) of β-D-fructose. This is shown in Fig. 3-27.

Though sucrose is readily digested by the body, it causes a rapid rise in blood glucose levels and is therefore not considered an ideal food source.

Disaccharide molecules, just like monosaccharides, can be α- and β-anomers, but when α or β is included in the name of the resulting disaccharide, this does not refer to the individual constituents that form the molecule, it refers to the particular

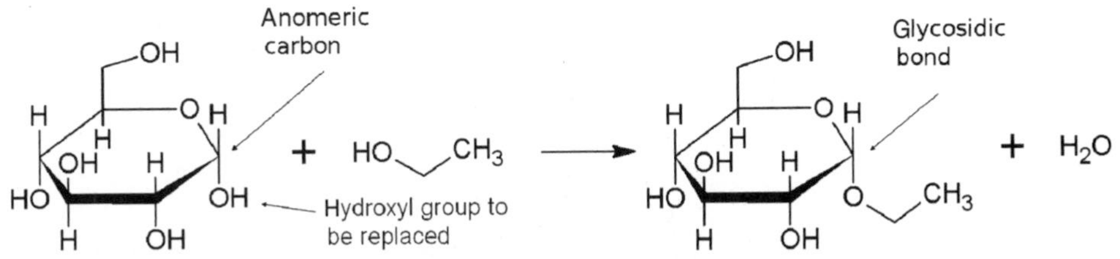

Figure 3-26   The reaction glucose + ethyl alcohol yields ethyl glucoside plus water.

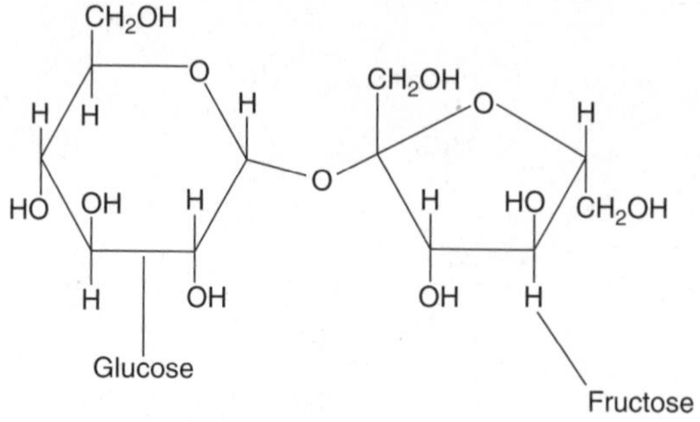

Figure 3-27   Sucrose is a 1,2-linkage between glucose and fructose.

anomer of the product. For example, *maltose* is a disaccharide formed from two α-D-glucose molecules, linked together via an α-1,4-glycosidic bond. This means that one glucose is linked through the anomeric carbon to the C-4 atom of the other glucose molecule. This bond can yield different anomers of maltose: α-D-maltose, β-D-maltose, or an open-chain form of maltose.

     Lactose is a sugar found in mammalian milk that is a disaccharide formed from galactose and glucose. The linkage that forms lactose is a β-1,4 linkage and is formed between C-1 of galactose and C-4 of glucose. See Fig. 3-28.

Polysaccharides

Polysaccharides are compounds containing many sugars linked together into large molecules. One of the primary functions of polysaccharides is energy storage or food reserve. Starch, found in plants, is a polymer actually made up of two more fundamental types of molecules called *amylopectin* and *amylose*. Starch is between

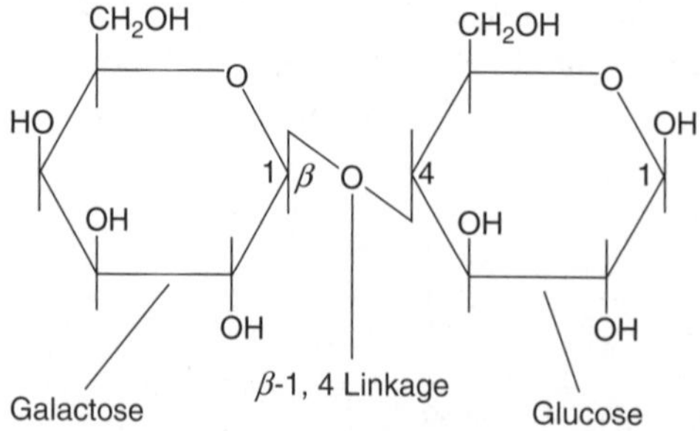

Figure 3-28   Lactose is a disaccharide formed from a β-1,4 linkage between β-galactose and glucose.

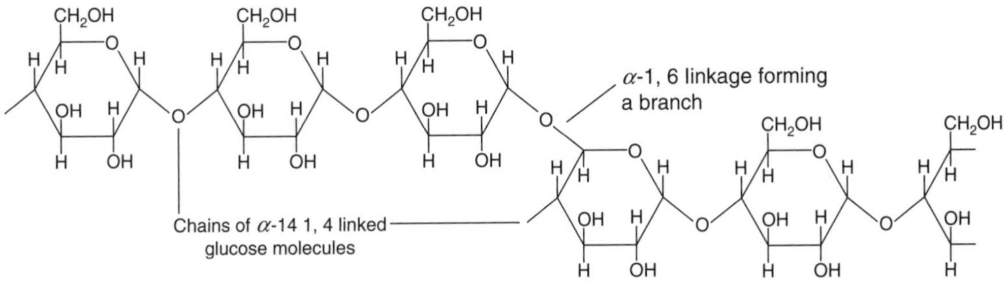

Figure 3-29   Amylopectin is a branched polymer made up of glucose units.

80 and 90% amylopectin, which is a branched polymer that consists of glucose units bonded together by α-1,4 linkages. These form chains that are connected to other chains of glucose units bonded together by α-1,4 linkages through an α-1,6 bond (which forms the branching). Amylopectin is water soluble. It is a very large molecule, and can contain up to 200,000 glucose units. Branching provided by the α-1,6 bond occurs every 24 to 30 glucose units. Amylopectin is shown in Fig. 3-29, where two chains of glucose units bonded together by α-1,4 linkages are connected by a branch provided by an α-1,6 bond.

Amylose is a linear polymer of glucose units linked together by α-1,4-glycosidic bonds. Since it is linear and without branching, it has a simpler structure than amylopectin. The number of glucose units in an amylose molecule can range into the thousands; typical numbers are between 100 and 3000 glucose units. Despite its simpler structure, amylose is actually harder to digest than amylopectin. However, lower forms of plants contain more amylose because it is easier to store than amylopectin. This is due to the fact that it forms simple linear chains. Higher plants utilize amylopectin for storage instead. Amylose is illustrated in Fig. 3-30.

Amylose and amylopectin are broken down in the body by enzymes. Human saliva and plants contain an enzyme called *α-amylase* that hydrolyzes the α-1,4 linkages in amylose and amylopectin.

Cellulose is another major polymer of glucose found in plants, where it is the primary component of cell walls giving plants their structure. It is a linear polymer

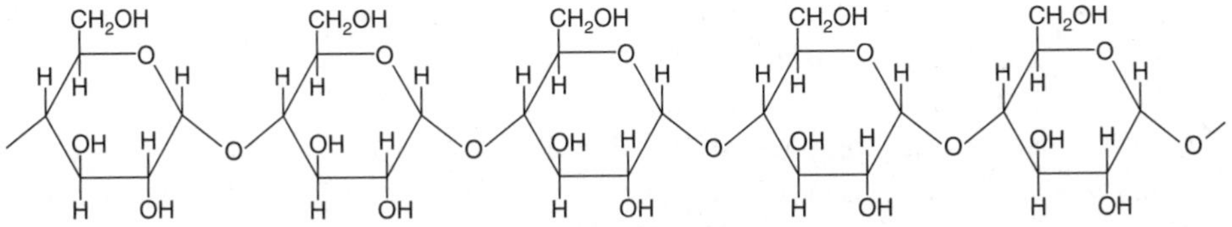

Figure 3-30   Amylose is a linear polymer of glucose molecules.

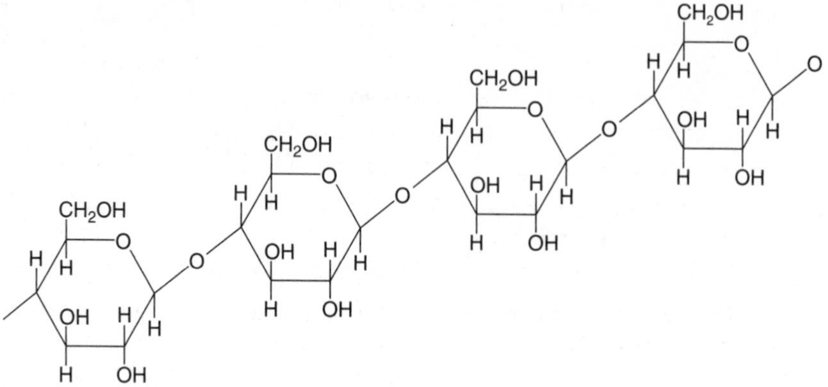

Figure 3-31   Cellulose is composed of glucose units linked together with β-1,4 bonds.

of glucose units linked together by β-1,4-glycosidic bonds. Cellulose is illustrated in Fig. 3-31.

Animals also store carbohydrates as polymers. The polysaccharide used for storage in animals is called *glycogen*. The structure of glycogen is similar to that of amylopectin, with branches occurring every 8 to 12 glucose units. In total, a glycogen molecule consists of 60,000 glucose units. Glycogen is constructed in the liver from individual glucose molecules. It is stored primarily in the liver but can also be found in muscle cells and in small amounts in other types of cells.

Glycoconjugates

A *glycoconjugate* is a carbohydrate (*glycan*) covalently linked with another compound. We only briefly summarize them here. The first type of glycoconjugate we consider are *glycoproteins,* which are proteins with monosaccharides, disaccharides, oligosaccharides, or polysaccharides (or a carbohydrate derivative) attached. These include carbohydrate chains that are between 1 and 30 base units in length. If the carbohydrate is bonded to the –OH group of serine or threonine, we say that the molecule is *O-linked*. If it is linked to an asparagine, then we say it is *N-linked*. Glycoproteins play a role in many important biological processes in eukaryotic cells. For example, they can form the basis of structural proteins, enzymes, or hormones.

A *proteoglycan* is a complex molecule consisting of polysaccharides together with *glucosaminoglycans* and proteins. A glucosaminoglycan is an unbranched polysaccharide with alternating hexosamine and uronic acid residues. These substances form a gel-like matrix and are used in cartilage and synovial fluid. Examples include chondroitin-6-sulfate and hyaluronic acid.

Summary

Carbohydrates are the most abundant biomolecule playing diverse roles in nature including structure, playing a role in DNA, and, most importantly, forming the basis of energy storage and metabolism. Large carbohydrates are constructed from fundamental units called monosaccharides. Monosaccharides come in different isomers, the most important of which are D and L isomers that react differently to light. Most biomolecules are D isomers but it is possible to have biologically active L isomers of some compounds. The most important monosaccharide in biological processes is glucose.

More complex molecules can be constructed by linking monosaccharides together. Two monosaccharides can be joined together with a glycosidic bond. The most familiar such compound is sucrose, which consists of a glucose and fructose molecule joined together. Polysaccharides are large carbohydrates built out of glucose molecules. Plants use starch, which consists of amylose and amylopectin, to store energy. Animals also store energy in the form of carbohydrates in glycogen. Cellulose is a long-chain molecule of glucose units that gives plants their structure.

Derivatives of carbohydrates also play an important role in biological processes. For example, a deoxycarbohydrate is derived from a carbohydrate by the removal of an oxygen. Ribose and deoxyribose are components of RNA and DNA, respectively.

Carbohydrates can form complex molecules with other constituents such as proteins. When this occurs, the resulting compound is called a glycoconjugate.

Quiz

1. Consider the molecules *psicose* and fructose shown in Fig. 3-32. These molecules are

   (a) Both D-isomers. Psicose is an enantiomer of fructose.

   (b) Both D-isomers. Psicose and fructose are diastereomers.

   (c) Psicose is the L-isomer configuration of fructose.

   (d) Psicose and fructose are not stereoisomers.

```
      CH₂OH                CH₂OH
       |                    |
       C═                   C═O
       |                    |
    H─C─OH             OH─C─H
       |                    |
    H─C─OH              H─C─OH
       |                    |
    H─C─OH              H─C─OH
       |                    |
      CH₂OH                CH₂OH
     Psicose              Fructose
```

Figure 3-32   Psicose and fructose are isomers.

2. Glucose and fructose

   (a) are diastereomers.

   (b) are not related at all.

   (c) are not isomers. Glucose is an aldehyde and fructose is a ketone.

   (d) are enantiomers.

3. A sugar alcohol is a compound that

   (a) is formed by reduction of glucose, replacing the aldehyde with an additional hydroxyl group.

   (b) is an enantiomer to a true alcohol.

   (c) only comes in L-isomers, and hence cannot be digested.

   (d) cannot be digested by oral bacteria and are diastereoisomers to glucose.

4. When the carboxyl group of a monosaccharide is reduced the result is a

   (a) deoxycarbohydrate.

   (b) reduced carbohydrate.

   (c) furanose ring.

   (d) sugar alcohol.

5. Two compounds are anomers if

   (a) their configurations differ only at the penultimate carbon.

   (b) their configurations differ at the hemiacetal or hemiketal carbon.

   (c) they exist in five- and six-membered ring configurations.

   (d) they are stereoisomers but are not mirror images of each other.

6. Which of the following alterations could be applied to maltose without changing the molecule into another compound?

   (a) Changing the configuration of the free anomeric hydroxyl group from α to β.

   (b) Changing the α-1,4 glycosidic linkage to a β-1,4-glycosidic linkage.

   (c) Neither (a) nor (b) is correct.

7. Which of the following is not a true statement about sorbitol?

   (a) Sorbitol is one member of a general class of molecules called sugar alcohols.

   (b) Sorbitol is derived from glucose.

   (c) To form sorbitol, an aldehyde is converted to a hydroxyl group.

(d) Sorbitol cannot be completely absorbed by the small intestine.

   (e) None of the above is false.

8. In what way does amylose differ from amylopectin?

   (a) It is a linear chain molecule composed of glucose units linked by α-1,6 bonds.

   (b) It is a branching molecule composed of glucose units linked by α-1,4 bonds.

   (c) It does not have branching.

   (d) It does have branching.

   (e) It is much larger than amylopectin.

9. To obtain the L-isomer of a D molecule,

   (a) change all groups connected to the penultimate carbon.

   (b) change all groups connected to the anomeric carbon.

   (c) change the anomeric hydroxyl group so that it is above the plane of the molecule.

   (d) change all groups connected to chiral carbons.

10. When it comes to biological systems

   (a) all biomolecules are D-isomers.

   (b) no biomolecules are D-isomers.

   (c) most biomolecules are D-isomers, but some bioactive L-isomers exist.

   (d) None of the above are true.

CHAPTER 4

Lipids

Lipids are a class of compounds distinguished by their *insolubility* in water and *solubility* in nonpolar solvents. Lipids are important in biological systems because they form the *cell membrane,* a mechanical barrier that divides a cell from the external environment. Lipids also provide energy for life and several essential vitamins are lipids. Lipids can be divided in two major classes, *nonsaponifiable lipids* and *saponifiable lipids.* A nonsaponifiable lipid cannot be broken up into smaller molecules by hydrolysis, but a saponifiable lipid contains one or more ester groups allowing it to undergo hydrolysis in the presence of an acid, base, or enzyme. Within these two major classes of lipids, there are several specific types of lipids important to life, including *fatty acids, triglycerides, glycerophospholipids, sphingolipids*, and *steroids*. Saponifiable lipids include *triglycerides, waxes, phospholipids*, and *sphingolipids*. Each of these categories can be further broken down. Nonsaponifiable lipids include steroids, prostaglandins, and terpenes.

Nonpolar lipids, such as triglycerides, are used for energy storage and fuel. Polar lipids, which can form a barrier with an external water environment, are used in membranes. Polar lipids include glycerophospholipids and sphingolipids. Fatty acids are important components of all of these lipids.

We now consider several lipids in detail.

Fatty Acids

A *fatty acid* is a molecule characterized by the presence of a carboxyl group attached to a long hydrocarbon chain. Therefore these are molecules with a formula R–COOH where R is a hydrocarbon chain. Fatty acids can be said to be carboxylic acids, and come in two major varieties.

- **Saturated fatty acids** This is a fatty acid that does not have any double bonds. We say that a fatty acid is saturated when every carbon atom in the hydrocarbon chain is bonded to as many hydrogen atoms as possible (the carbon atoms are saturated with hydrogen). Saturated fatty acids are solids at room temperature. Animal fats are a source of saturated fatty acids. In addition, fatty acids pack easily and form rigid structures (e.g., fatty acids are found in membranes).

- **Unsaturated fatty acids** An unsaturated fatty acid can have one or more double bonds along its hydrocarbon chain. A fatty acid with one double bond is called *monounsaturated.* If it contains two or more double bonds, we say that the fatty acid is *polyunsaturated.* Fatty acids only contain cis double bonds (see later for a discussion of *cis* and *trans*). The melting point of a fatty acid is influenced by the number of double bonds that the molecule contains and by the length of the hydrocarbon tail. The more double bonds it contains, the lower the melting point. As the length of the tail increases, the melting point increases. The melting point *decreases* as the number of double bonds *increases* because of the cis geometry of the double bonds. This introduces kinks in the hydrocarbon chain that *decreases* the number of van der Waals interactions. Conversely, as the length of the tail is *increased*, the number of van der Waals interactions is *increased*, raising the melting point. Obviously only the latter factor will affect the melting point of saturated fatty acids. Unsaturated fatty acids are liquids at room temperature. In the solid state, unsaturated fatty acids do not pack as well as straight-chained saturated fatty acids. As a result, intermolecular forces between molecules are lower, and it is easier to separate the molecules by raising the temperature, resulting in a lower melting point. Plants are the source of unsaturated fatty acids.

Common saturated fatty acids contain between 12 and 20 carbon atoms. Table 4-1 lists some common saturated fatty acids. Note the increase in melting point with the increase in the length of the hydrocarbon chain.

The straight-chain structure of saturated fatty acids is illustrated in Fig. 4-1 which displays lauric acid.

In Fig. 4-2, we show an illustration of oleic acid, an unsaturated fatty acid, which shows the prominent kink in the molecule.

Table 4-1   Common Saturated Fatty Acids

Fatty Acid	Carbon Atoms	Formula	Melting Point (°C)
Lauric	12	$CH_3(CH_2)_{10}COOH$	44
Myristic	14	$CH_3(CH_2)_{12}COOH$	54
Palmitic	16	$CH_3(CH_2)_{14}COOH$	63
Stearic	18	$CH_3(CH_2)_{16}COOH$	70
Arachidic	20	$CH_3(CH_2)_{18}COOH$	77

Some common unsaturated fatty acids are listed in Table 4-2. Notice that this table illustrates the melting point of a fatty acid as it relates to its characteristics. Oleic acid and palmitoleic acid have the same number of double bonds, but oleic acid has a longer hydrocarbon chain. Hence the melting point is increased. Now compare the linoleic and linolenic fatty acids. They both have the same number of carbon atoms, hence their hydrocarbon chains are the same length. But linolenic acid has *more* double bonds, therefore its melting point is *lower*.

OMEGA FATTY ACIDS

Unsaturated fatty acids can also be classified according to the location of the closest double bond to the methyl end of the carbon chain furthest from the carboxyl group. This is done by specifying an *omega number* for the fatty acid. Looking at Table 4-2, we examine the formula of the fatty acid and count the number of carbon atoms starting with the CH_3 on the left up to the first double bond. Hence

- Palmitoleic acid is an omega-7 fatty acid.
- Oleic acid is an omega-9 fatty acid.
- Linoleic acid is an omega-6 fatty acid.
- Linolenic acid is an omega-3 fatty acid.

Figure 4-1   A rendering of lauric acid. This is a saturated fatty acid which has a straight hydrocarbon chain.

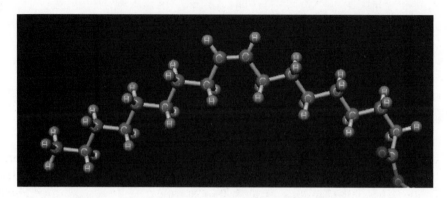

Figure 4-2   A rendering of oleic acid, an unsaturated fatty acid. Note the kink in the molecule, as opposed to the straight-chain nature of the saturated fatty acid shown in Fig. 4-1.

Omega-3 and omega-6 fatty acids are believed to be very important for cardiovascular health. Linolenic acid is an essential fatty acid because other omega-3 fatty acids needed by the human body are synthesized from it.

CIS- AND TRANS-FATTY ACIDS

Naturally occurring fatty acids have cis bonds. *Trans-fatty acids* are created artificially using a process called *hydrogenation.* A trans-fatty acid has a trans configuration rather than cis configuration at each double bond. This causes the molecule to straighten. These two stereoisomers can be distinguished in the following way:

- In a *cis* stereoisomer, two similar groups attached to the carbon double bond are found on the same side.

- In a *trans* stereoisomer, two similar groups attached to the carbon double bond are found on opposite sides.

Table 4-2   Common Unsaturated Fatty Acids

Fatty Acid	Carbon Atoms	Formula	Melting Point (°C)
Palmitoleic	16	$CH_3(CH_2)_5CH=CH(CH_2)_7COOH$	1
Oleic	18	$CH_3(CH_2)_7CH=CH(CH_2)_7COOH$	13
Linoleic	18	$CH_3(CH_2)_4(CH=CHCH_2)_2(CH_2)_6COOH$	−5
Linolenic	18	$CH_3CH_2(CH=CHCH_2)_3(CH_2)_6COOH$	−11
Arachidonic	20	$CH_3(CH_2)_4(CH=CHCH_2)_4(CH_2)_2COOH$	−49

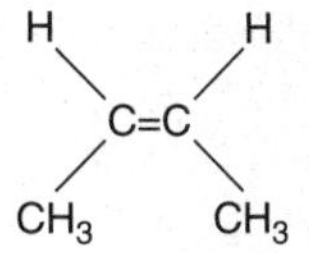

H   H
\\   /
C=C
/   \\
CH₃   CH₃

Figure 4-3   The cis configuration of 2-butene has the methyl groups on the *same* side
of the double bond.

The designation of a stereoisomer as cis or trans is a geometrical designation. The atoms are arranged differently in space and there is a high energy barrier between the two configurations. As a result, two compounds that are differentiated as cis or trans are *different compounds.* We can understand the difference between cis and trans configurations by considering a simple example. Butene comes in two varieties, shown in Figs. 4-3 and 4-4.

Trans bonds, unlike cis bonds, do not introduce kinks in the hydrocarbon chain of a fatty acid. As a result, the effect of double bonds (in lowering the melting point) does not occur in the trans isomer of a given fatty acid. Trans-fatty acids behave more like saturated fatty acids, so the melting point is dependent on the length of the hydrocarbon chain. Trans-fatty acids have much higher melting points than the corresponding unsaturated cis-fatty acid. Although they are created by saturating vegetable oils, trans-fatty acids are considered unhealthy and are believed to lead to cardiovascular disease if consumed in large amounts.

The key features of a trans-fatty acid are that the hydrocarbon chain contains a double bond, but the chain is straight instead of containing kinks like an unsaturated fatty acid. A trans-fatty acid is shown in Fig. 4-5.

ESSENTIAL AND NONESSENTIAL FATTY ACIDS AND THEIR ROLE IN DIET

If a fatty acid can only be obtained from the diet (for humans), we say that the fatty acid is an *essential fatty acid.* Two fatty acids cannot be synthesized in the human body and are therefore essential. These are *linoleic* and *linolenic* fatty acids, which

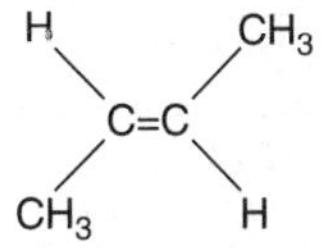

H   CH₃
\\   /
C=C
/   \\
CH₃   H

Figure 4-4   In the trans configuration of 2-butene, the methyl groups are on the *opposite*
side of the double bond.

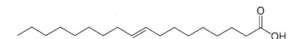

Figure 4-5   An example of a trans-fatty acid, *elaidic acid.* Note that although it has a double bond, since this is a trans isomer the hydrocarbon chain is straight.

are both unsaturated. Nonessential fatty acids can be made by the human body and so do not need to be obtained from diet alone. These are made from carbohydrates and proteins or from other fatty acids.

Fatty acids are an important source of energy. While carbohydrates or proteins only provide 4 kcal/g, fatty acids provide more than twice the energy per unit weight at 9 kcal/g. This is one reason why a high-fat diet can lead to obesity.

Triglycerides

A triglyceride (often called *tryglycerol*—we will use both terms interchangeably) is a fatty acid trimester of glycerol, which is illustrated in Fig. 4-6. Triglycerides are important for human health in that they provide most of the lipids in our diet. Notice that glycerol has three hydroxyl groups. Fatty acids can be attached at these three sites forming a triglyceride.

One important characteristic of a tryglycerol is its state at room temperature. The degree of saturation and the length of their chains attached to the glycerol backbone both determine their state at room temperature. So we see that not surprisingly, the state of the tryglycerol is determined by how the fatty acid chains it contains behave. Hence

- Short-chain unsaturated triglycerides are *liquid* at room temperature.
- Long-chain saturated triglycerides are *solid* at room temperature.

Animal fats contain a high amount of saturated triglycerides while plant oils contain a high amount of unsaturated triglycerides. Think *lard* when thinking about a saturated triglyceride, and think *vegetable oil* when thinking about an unsaturated triglyceride. While neither is healthy when consumed in excess, vegetable oils are far healthier than lard, so a biochemist can keep in the back of their mind that fats that are solid at room temperature are not as healthy as those that are liquid.

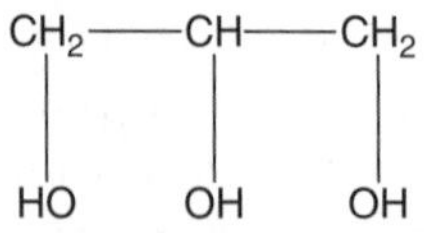

$$CH_2 \text{——} CH \text{——} CH_2$$

HO   OH   OH

Figure 4-6   An illustration of a glycerol molecule.

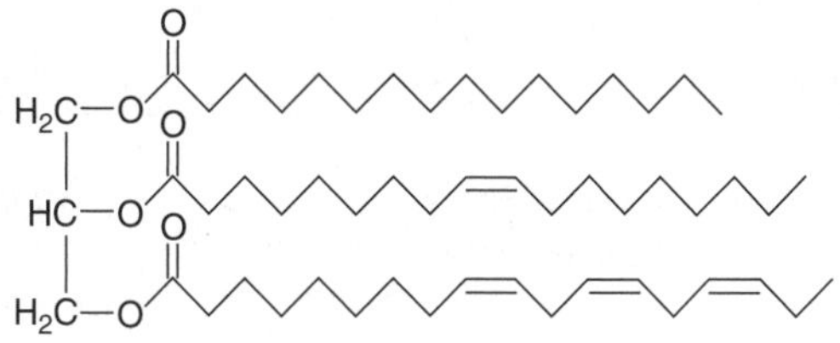

Figure 4-7   A triglyceride.

THE ROLE OF TRIGLYCERIDES IN HEALTH

While fat may seem bad, triglycerides play many important roles in the body. For example, triglycerides can be used for energy storage in animals. This food reserve can be called upon during periods of starvation, with the high-calorie content of the fatty acids adding to the value of storing fat and providing much needed energy. In addition, triglycerides can provide insulation for animals in the form of body fat, which allows them to survive in colder temperatures. These two roles played by fat in the body, which arose over eons of evolution, are now deemed undesirable in modern industrialized society where humans no longer face starvation or have to deal directly with cold weather.

THE FORMATION OF TRYGLYCEROLS

Trigylcerides are formed when each of the OH groups in glycerol reacts with the COOH group of a fatty acid to create an ester group. Three water molecules are liberated in the process. The resulting molecule is illustrated in general in Fig. 4-7. Note the fatty acid chains which extend from the glycerol backbone.

   If all three R groups are the same, that is, the fatty acid that leads to the formation of each ester group in the resulting tryglycerol is the same, we say that the compound is a *simple tryglycerol.* In nature, the R groups will be different. In that case we call the resulting compound a *complex tryglycerol.*

Sphingolipids

A *sphingolipid* is an important constituent of the cell membrane which is based on a backbone molecule called *sphingosine* rather than glycerol. Sphingolipids can be found throughout the nervous systems of mammals, where they form a component of the *myelin sheath,* which is a fatty layer that provides insulation for the axons of neurons.

   Components can be incorporated with a sphingosine molecule via reactions at the NH_2 and OH groups. For example, to obtain a *sphingomyelin,* a fatty acid is attached at the location of the NH_2 group of sphingosine. Another type of sphingolipid called a cerebroside has a saccharide unit attached at the location of the OH group

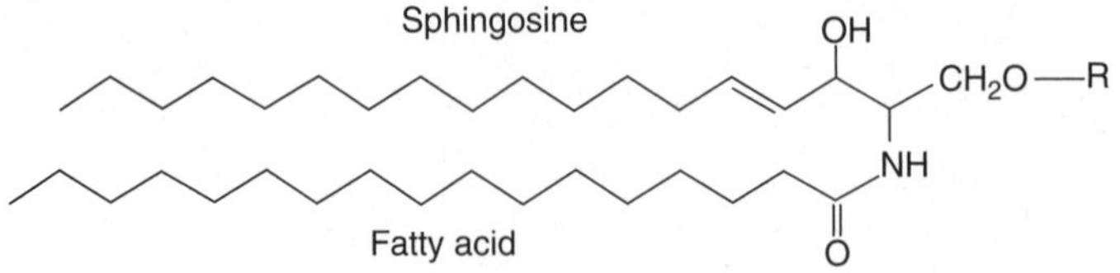

Sphingosine

OH

CH₂O—R

NH

Fatty acid

O

Figure 4-8 A sphingolipid consists of a fatty acid bound to sphingosine and an R group.
(*Image courtesy of Wikipedia*)

of sphingosine. Cerebrosides are also abundant in the nervous system. An illustration of a sphingolipid is shown in Fig. 4-8.

The structure of a sphingolipid can be described as consisting of an unbranched 18-carbon alcohol with one trans double bond between carbons 4 and 5. These compounds have an NH_3^+ group bonded to the second carbon, and hydroxyl groups located at carbons 1 and 3. Besides playing a role in the central nervous system, it is believed that sphingolipids function as cell surface markers, providing ABO blood type antigens, for example.

Nonsaponifiable Lipids

The defining characteristic of a lipid is that it is insoluble in water and it is soluble in a nonpolar solvent. A nonsaponifiable lipid is one that cannot be broken down by hydrolysis. These characteristics bring several important biomolecules under the lipid umbrella that you may not think of as "fats." We begin our discussion of nonsaponifiable lipids by taking a brief look at *steroids*.

A *steroid* is a biologically important lipid that cannot be broken down into smaller molecules by the process of hydrolysis because it lacks ester groups. The defining characteristic of a steroid is the presence of a four-ring system that gives it structure. This ring system contains a single five-membered carbon ring together with three six-membered carbon rings.

The four-ring structure characteristic of steroids is illustrated in Fig. 4-9.

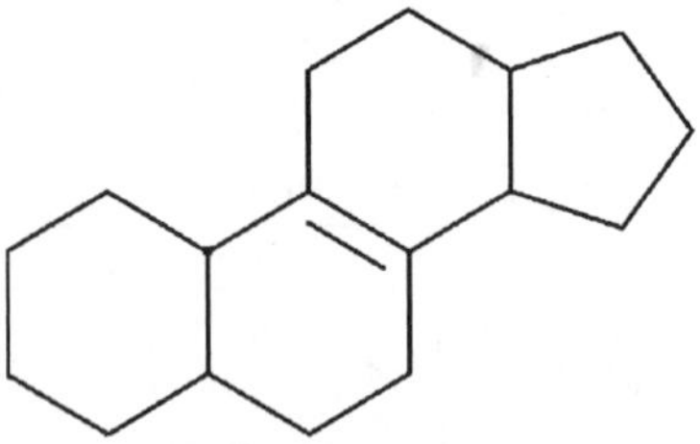

Figure 4-9 A steroid has a four-ring structure. Note that three of the rings have six members, and the remaining ring, found on the right, has five members.

Perhaps the most fundamental and famous steroid is cholesterol, which we discuss in the next section. Here we discuss several other steroids that are important in the body. These include the following:

- **Androgens** These are "male sex hormones" that regulate the development of the male reproductive system and the secondary sexual characteristics in males.
- **Progesterone, estrone, and estradiol** These are "female sex hormones" that regulate the development of the female reproductive system and are responsible for the maintenance of secondary sexual characteristics in females.
- **Aldosterone** This steroid controls water and electrolyte balances.
- **Cortisone** This compound is involved in metabolism and in controlling inflammation.
- **Bile salts** Facilitates the digestion of certain lipids and the absorption of fat-soluble vitamins.
- **Vitamin D** An important steroid that controls calcium absorption and deposition in the bone. Recent research also suggests that vitamin D plays a fundamental role in the prevention of many cancers. High consumption of vitamin D and sun exposure appear to reduce cancer risk.

Prostaglandins are nonsaponifiable lipids that are involved in several body functions. They consist of a 20-carbon chain that includes a five-membered ring at the end. One of the most important roles of prostaglandins is in the regulation of blood pressure. They also control blood clotting and induce labor.

Terpenes are large molecules constructed out of an *isoprene,* which is a branched carbon-5 unit (see Fig. 4-10). You know terpenes best as several common vitamins. Some biologically important terpenes include

- **Vitamin A** Important for healthy vision, in particular night vision
- **Vitamin E** An important antioxidant that is involved in the maintenance of cell membrane integrity
- **Vitamin K** Involved in blood clotting

An example of a terpene is shown in Fig. 4-11, which displays vitamin K.

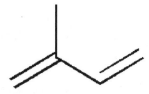

Figure 4-10 An isoprene is a five-membered carbon branch. Isoprenes can be assembled into vitamins called terpenes.

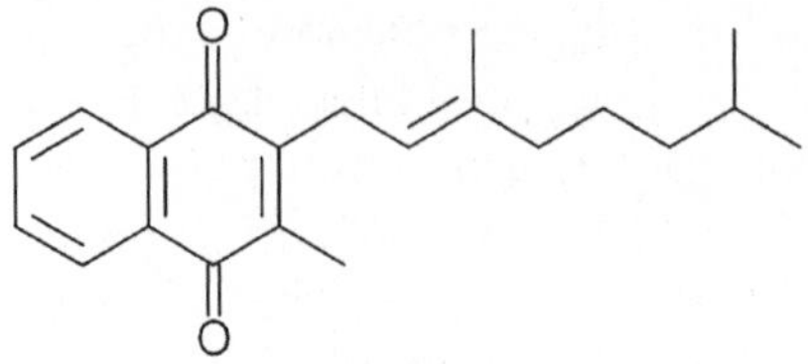

Figure 4-11   An example of a terpene, a variant of vitamin K.

Cholesterol

Cholesterol is an important lipid found in the cell membrane. It is a *sterol*, which means that cholesterol is a combination of a steroid and an alcohol. It is an important component of cell membranes and is also the basis for the synthesis of other steroids, including the sex hormones estradiol and testosterone, as well as other steroids such as cortisone and vitamin D. An illustration of cholesterol is shown in Fig. 4-12.

In the cell membrane, the steroid ring structure of cholesterol provides a rigid hydrophobic structure that helps boost the rigidity of the cell membrane. Without cholesterol the cell membrane would be too fluid.

In the human body, cholesterol is synthesized in the liver. Cholesterol is insoluble in the blood, so when it is released into the blood stream it forms complexes with *lipoproteins*. Cholesterol can bind to two types of lipoprotein, called *high-density lipoprotein (HDL)* and *low-density lipoprotein (LDL)*. A lipoprotein is a spherical molecule with water soluble proteins on the exterior. Therefore, when cholesterol is bound to a lipoprotein, it becomes blood soluble and can be transported throughout the body.

HDL cholesterol is transported back to the liver. If HDL levels are low, then the blood level of cholesterol will increase. High levels of blood cholesterol are associated with plaque formation in the arteries, which can lead to heart disease and stroke.

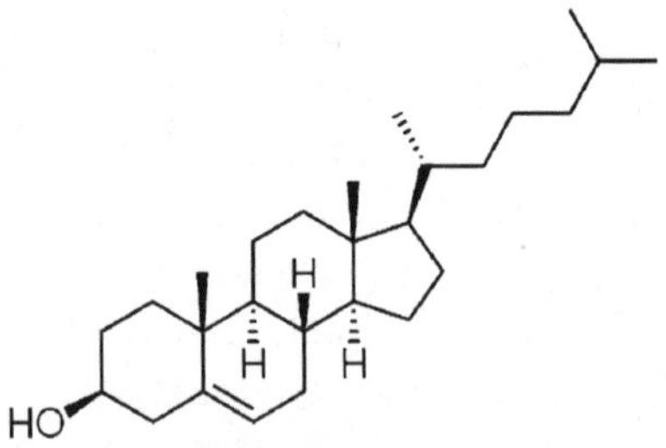

Figure 4-12   Cholesterol, which consists of an OH bound to a steroid ring formation and a hydrocarbon chain.

While most cholesterol in the body is synthesized in the liver, dietary cholesterol also adds to the total blood levels. Cholesterol intake from the diet enters the bloodstream in the LDL form. This helps explain why consumption of foods with high-cholesterol content can lead to increased blood levels of cholesterol which is bad for health. So reducing the cholesterol in the diet can lower the blood level of cholesterol. This can reduce the amount of plaque formation. Aerobic exercise also contributes to health by increasing HDL levels in the blood. Hence more cholesterol is returned to the liver leading to a lower blood level of cholesterol, and reduced plaque formation.

CHOLESTEROL'S ROLE IN THE CELL MEMBRANE

A cholesterol molecule has a hydroxyl group which acts to bind it to phospholipids in the cell membrane. Meanwhile, the steroid portion of the molecule acts as an anchor of sorts, interacting with the fatty acid chains of phospholipids. It is embedded directly in the cell membrane, helping guard it from too much fluidity.

An Overview of the Cell Membrane

A *cell membrane* is a structure constructed out of lipids, cholesterol, proteins, and carbohydrates that divides the cytoplasm and intracellular components of a cell from the external environment. The primary characteristic of the cell membrane, whose main component is constructed out of lipids, is that it consists of a *lipid bilayer* which functions to keep the external aqueous environment out and internal water in. You can think of a cell membrane by analogy as a plastic bag.

The physical division that the cell membrane provides is accomplished with phospholipids that have a hydrophilic head and hydrophobic tails. The outer layer of the cell membrane is formed by closely packed phospholipids, with the hydrophilic head facing outward, into the external watery environment. The tails, which are hydrophobic, extend downward from the heads inward where they bond to the tails of other phospholipids that form the inner side of the lipid bilayer. Both external layers of the cell membrane are water friendly and hence are in contact with aqueous solutions. When a molecule contains both a hydrophilic and a hydrophobic component, we say that it is *amphipathic*. An example of a lipid membrane is shown in Fig. 4-13—this is a single lipid layer showing the hydrophilic heads and the hydrophobic tails.

When placed in an aqueous environment, amphipathic lipids will spontaneously arrange themselves into a spherical lipid bilayer structure. This arrangement is such that the hydrophobic tails are protected from the external aqueous environment, while the hydrophilic heads face it, well head-on. The formation of the spherical structure traps some aqueous solution inside the lipid bilayer, providing an internal environment

Figure 4-13 A lipid layer consists of water-loving (hydrophilic) heads and water-hating (hydrophobic) tails. The tails extend below the heads which face a watery or aqueous environment. See Fig. 4-14.

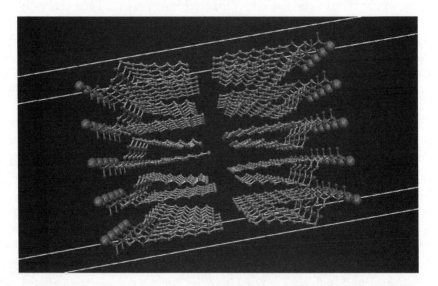

Figure 4-14 A lipid bilayer consists of two lipid layers of the type shown in Fig. 4-13. The hydrophobic tails join in the middle forming a bilayer-membrane structure.

faced by the internal layer of hydrophilic heads. Life may have gotten started by the spontaneous formation of lipid bilayer spheres, which may have trapped other chemicals such as RNA that led to the construction of more complicated proto-organisms.

The side of the lipid bilayer that faces the external environment is called the *extracellular side,* while the side of the lipid bilayer that faces the internal cell environment is called the *intracellular side.* This division of external and internal is what makes life possible, without it life would be a meaningless phenomenon. The cell membrane is the first component that gives identity to an organism by distinguishing or separating it from its environment. This allows metabolic processes to take place within the cell; it allows the cell to store genetic information in the form of RNA and DNA; and it allows the cell to maintain necessary pH and ionic composition inside its internal fluids. Moreover, the cell membrane functions to dispose of waste materials to the outside environment.

The specific lipid content in a cell membrane consists of *phosphoglycerides* (this is a phospholipid containing a glycerol phosphate), *plasmalogens* (a lipid with an ether-linked alkene at the first carbon of the glycerol), sphingomyelins, glycolipids, and cholesterol. A phospholipid has a phosphate ester group and ionic charges, with a fatty acid chain that contains an even number of carbon atoms which typically ranges between 14 and 24. The fatty acid chains can be saturated or unsaturated, with the unsaturated fatty acids adding to the flexibility of the cell membrane.

A cell membrane is a liquid-like structure that has a certain amount of flexibility. As mentioned above, that flexibility is provided by the presence of unsaturated fatty acids. If you recall from the opening section of this chapter, unsaturated fatty acids (in the natural state) have cis double bonds among one or more of their carbon atoms that add a kink to the molecule. When they are packed together or with other molecules, these kinks serve to prevent the lipid molecules in the biliary from getting too close to one another. The end result is that intermolecular interactions among components of the membrane are decreased, giving it a flexible structure.

Of course, you would not want the cell membrane to be too flexible, otherwise it would just fall apart. Stability can be added to the cell membrane via the presence of cholesterol, as mentioned in the previous section. The hydrophobic portion of a cholesterol molecule is its four-ring system which is rigid as compared with the lipid bilayer. Adding cholesterol to the cell membrane keeps it stable by preventing it from being too fluid.

Proteins also play an important role in the cell membrane. A cell membrane can consist of anywhere between 20 and 75% protein. An *integral membrane protein* is one that is embedded directly in the lipid bilayer, sometimes extending from the extracellular side down into the intracellular side. Integral proteins act as a *gate* that can allow certain molecules (such as glucose) to enter the interior of the cell. More specifically, integral proteins function as ion channels and proton pumps. They may also act as receptors. They determine what comes in and what leaves a cell.

A *peripheral protein* is one that is associated with the cell membrane but that is not embedded directly in it. This association is usually temporary, and can involve an association with an integral protein. Peripheral proteins may be involved in some cellular process, for example, an enzyme might act as a peripheral protein driving some chemical reaction. Hormones can also act as peripheral proteins.

Molecules and ions need to be transported across the cellular membrane in order for the cell to function. This can be accomplished in one of the following ways:

- **Diffusion**   If a concentration gradient exists from inside to outside the cell or vice versa, relatively small hydrophobic molecules are able to diffuse across the cell membrane passing through the bilayer of hydrophobic tails. An example of this is O_2 diffusion.

- **Active transport**   Integral proteins can actively move molecules in and out of the cell against the concentration gradient, a process that requires energy input.

- **Passive transport**   This involves the movement of molecules along a concentration gradient, from high to low concentration. The movement of the molecules is done with the assistance of an integral protein which acts as a *channel*. Large polar uncharged molecules are transported across cell membranes via passive transport, such as glucose. Ions are also transported across cell membranes via passive transport. An example is the movement of ions across neural axons.

A complete picture of a cell membrane is shown in Fig. 4-15 which includes proteins, the lipid bilayer, and cholesterol.

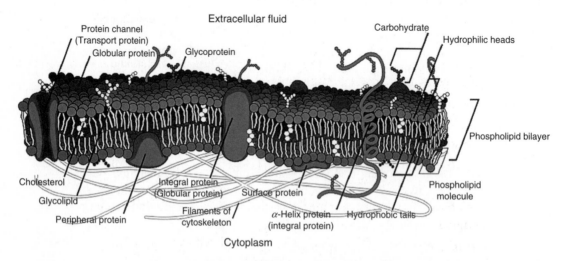

Figure 4-15   An illustration of the cell membrane, showing embedded proteins, cholesterol, and carbohydrate components. (*Image courtesy of Wikipedia*)

Summary

Lipids play an important role in many biological processes, provide a high-calorie source of energy and provide the structure of the cell membrane. The key component of many lipids are the fatty acids, which can be saturated, meaning that they include a long hydrocarbon chain with the maximum number of hydrogen atoms bonded to each carbon, or unsaturated meaning that the chain contains one or more double bonds. Naturally occurring fatty acids have cis double bonds, but trans double bonds can be created using hydrogenation. Triglycerides are molecules consisting of a glycerol backbone connected to fatty acid chains.

Steroids are molecules that are classified as nonsaponifiable lipids because they cannot undergo hydrolysis. They consist of a four-ring system. Important steroids include the sex hormones androgens and estrogens, along with cortisone and vitamin D. Cholesterol, an important steroid derivative, is important in the structure of cell membranes and also influences cardiovascular health.

Quiz

1. In an unsaturated fatty acid

   (a) melting point is increased by the length of the hydrocarbon tail and is increased by the degree of unsaturation.

   (b) melting point is decreased by the length of the hydrocarbon tail and is increased by the degree of unsaturation.

   (c) melting point is increased by the length of the hydrocarbon tail and is decreased by the degree of unsaturation.

   (d) melting point is decreased by the length of the hydrocarbon tail and is decreased by the degree of unsaturation.

2. Which of the following statements about trans fatty acids is false?

   (a) Trans-fatty acids have straight hydrocarbon chains, while the cis variety has one or more kinks.

   (b) Trans-fatty acids can be derived from plant oils using a process called hydrogenation.

   (c) Trans-fatty acids have lower melting points than the corresponding cis configuration.

   (d) A trans stereoisomer differs from a cis stereoisomer in that similar groups attached to the carbon double bond are found on opposite sides.

   (e) None of the above is false.

3. Consider *erucic acid* whose formula is $CH_3(CH_2)_7CH{=}CH(CH_2)_{11}COOH$. It is

   (a) an omega-9 fatty acid.

   (b) an unsaturated fatty acid.

   (c) an omega-7 unsaturated fatty acid.

   (d) an omega-9 saturated fatty acid.

   (e) Both (a) and (b) are correct.

4. Nonessential fatty acids

   (a) include the groups linoleic and linolenic fatty acids.

   (b) can be synthesized by the human body, so do not need to be directly consumed in the diet.

   (c) are not required by the human body, so do not need to be directly consumed in the diet.

   (d) include only saturated fatty acids.

5. Which of the following is not a true statement about triglycerides.

   (a) Short chain, unsaturated triglycerides are liquid at room temperature.

   (b) Long chain, saturated triglycerides tend to be solid at room temperature.

   (c) Short chain, saturated triglycerides are solid at room temperature.

   (d) Triglycerides are formed with a glycerol backbone and two or three fatty acids. They can be solid or liquid at room temperature depending on the length of the chains and the degree of saturation.

6. Which of the following statements about steroids is false?

   (a) A steroid is a nonsaponifiable lipid meaning that it cannot be broken down by hydrolysis.

   (b) A steroid contains a four-ring system.

   (c) Steroids are important regulators of may bodily functions including the regulation of secondary sexual characteristics.

   (d) A steroid is a nonsaponifiable lipid meaning that it can be broken down by hydrolysis.

7. Dietary cholesterol can be an important factor in heart health because

   (a) it is not absorbed by the liver.

   (b) it enters the blood stream bound to a low-density lipoprotein, so can raise blood cholesterol levels.

(c) it enters the blood stream bound to a high-density lipoprotein, so it can raise blood cholesterol levels.

(d) it is absorbed in the small intestine.

8. The structure of cholesterol is best described as

   (a) a five-ring system, bound to a hydroxyl group.

   (b) a four-ring system, bound to a hydroxyl group.

   (c) a four-ring system with one five-membered ring, bound to a hydroxyl group and a hydrocarbon chain.

   (d) a four-ring system with two five-membered rings, bound to a hydroxyl group and a hydrocarbon chain.

9. A sphingolipid

   (a) is structurally similar to a tryglycerol, but has a sphingosine backbone instead of a glycerol backbone.

   (b) has two NH_2 groups.

   (c) is an important regulator in cholesterol metabolism.

   (d) None of the above.

10. The cell membrane

   (a) is made more flexible by the presence of unsaturated fatty acids because the kinks in the cis double bonds minimize close packing of molecules in the lipid bilayer.

   (b) is made more flexible by the presence of cholesterol, whose four-ring structure minimizes the close packing of molecules in the lipid bilayer.

   (c) is made less flexible by the presence of unsaturated fatty acids, whose presence increases intermolecular attraction among components in the lipid bilayer.

   (d) is made more flexible by the presence of unsaturated fatty acids because the kinks in the trans double bonds minimize close packing of molecules in the lipid bilayer.

   (e) None of the above is correct.

CHAPTER 5

Lipid Metabolism

Lipids, in the form of triaglycerol and lipoproteins among other biomolecules, are a major component of the diet. Triaglycerols, in particular, are an important calorie source that stores a large amount of energy. While metabolism of carbohydrates and proteins produces around 4 to 5 kcal/g, the metabolism of triaglycerols produces nearly twice that amount at 9 kcal/g. The digestion of fats requires the action of enzymes produced in the pancreas and *bile salts* produced by the gall bladder. In this chapter, we give a basic overview of lipid metabolism.

Lipases and Phospholipases

A *lipase* is a hydrolytic enzyme that works in the small intestine to aid in the digestion of lipids. In particular, a lipase acts on dietary triaglycerol by hydrolyzing ester bonds between a fatty acid and a glycerol. Lipases are produced in the pancreas.

To catalyze the cleavage of the molecule, the enzyme must work in the presence of water. Some key facts about lipid digestion are:

- Triaglycerols are water insoluble.
- Digestive enzymes are water soluble.
  - Therefore, digestive enzymes must act at water-lipid interfaces.

The products produced by the action of a lipase are:

- 2-Monoacylglycerol
- Two fatty acid molecules

The digestion process of triacylglycerol works as follows:

- Lipids are digested by lipase enzymes.
- The products of digestion become bound to *bile salt micelles* (see below).
- Fatty acids are absorbed from the micelles to cells in the small intestine.
- The fatty acids are transported to the liver.
- They are bound as lipoproteins that can be transported to cells to be used as needed.

Recall from our chapter on lipids that many fat molecules exist as *phospholipids*. A *phospholipase* is an enzyme which hydrolyzes a single ester bond of a phospholipids between a fatty acid and a glycerol. There are four types or classes of phospholipases. These are:

- Phospholipase A   This class has two subclasses. Phospholipase A1 hydrolyzes the ester bond between glycerol and a fatty acid at position 1 of the carbon chain of a phosphoglyceride, while phospholipase A2 hydrolyzes the ester bond between glycerol and a fatty acid at position 2 of the carbon chain of a phosphoglyceride.
- Phospholipase B hydrolyzes the ester bond at both positions 1 and 2 of a phosphoglyceride.
- Phospholipase C cleaves the molecule at the phosphate.
- Phospholipase D cleaves the molecule after the phosphate.

The points of action of each phospholipase enzyme are indicated in Fig. 5-1.

The cleavage of a lipid or phospholipid produces fatty acids and lysophosphoglycerides, respectively. These molecules act as *detergents* in the digestive system. That is, they act to break up large droplets of fat into small droplets which are easier to digest.

Bile salts aid in the digestive process by helping to increase the surface area of the water-lipid interface, and they act as detergents. Bile salts increase the surface

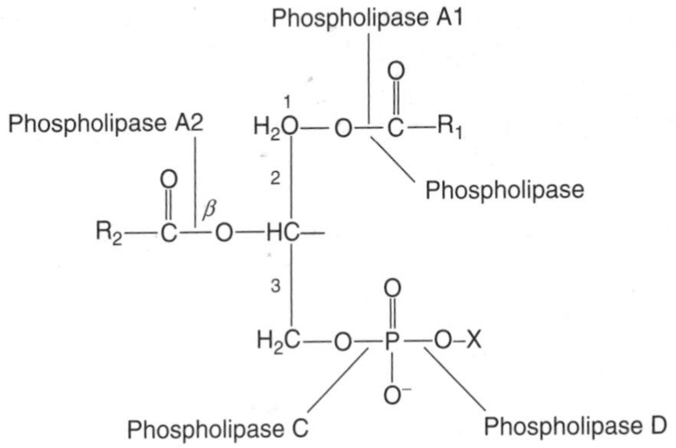

Phospholipase A1

Phospholipase A2 H_2C—O—$\overset{\overset{\displaystyle O}{\|}}{C}$—$R_1$

Phospholipase

R_2—$\overset{\overset{\displaystyle O}{\|}}{C}$—O—HC—

H_2C—O—$\overset{\overset{\displaystyle O}{\|}}{\underset{\underset{\displaystyle O^-}{|}}{P}}$—O-X

Phospholipase C   Phospholipase D

Figure 5-1   An illustration of where phospholipase enzymes act to cleave a phospholipid molecule.

area of the water-lipid interface by *emulsifying* the lipids, which means they are broken up into smaller droplets. So bile salts also act as detergents, which aid in the digestive process. Moreover, the bile salts form micelles which are single-layered structures (spherical in the case of bile salts) with the hydrophobic tails pointing inward. These micelles incorporate triacylglycerol, cholesterol, fat-soluble vitamins, fatty acids and 2-monoacylglycerol. When this happens we say that the micelle is a *mixed micelle*. These mixed micelles migrate to the intestinal wall where they release fatty acids, 2-monoacylglycerol, and fat-soluble vitamins that are absorbed by epithelial cells of the intestine. Bile salts are a derivative of cholesterol made in the liver and released by the gall bladder in the form of *bile* to the small intestine.

Bile salts are required for the absorption of fat-soluble vitamins. The fat-soluble vitamins that are digested in this way include:

- Vitamin A
- Vitamin D
- Vitamin E
- Vitamin K

LIPID TRANSPORT

Lipids are moved through the blood stream by a complex between a protein and a lipid called a *lipoprotein*. These molecules are classified by their density. Since lipids are low-density molecules, the lower the density of the lipoprotein, the higher the ratio of lipid to protein in the molecule. These molecules include:

- **Chylomicrons**   the lowest density lipoprotein
- **VLDL**   very low-density lipoproteins

- **LDL**　low-density lipoproteins
- **HDL**　high-density lipoproteins
- **IDL**　intermediate-density lipoproteins (derived from VLDLs in the formation of LDLs)

As noted above, chylomicrons are the lowest density lipoprotein. These molecules are composed primarily of triacylglycerols and are in fact 95% lipids. They serve to transport lipids for storage in adipose tissue. Once there, the chylomicrons are degraded and taken up by the liver.

Next on the scale we find the very low-density lipoproteins (VLDL). These molecules are between 90 and 95% fat. Like chylomicrons, they also act to transport triacylglycerols to adipose tissue. VLDLs are degraded into LDLs.

We have already "met" LDL and HDL lipoproteins in our discussion of cholesterol. The primary transport protein for cholesterol is LDL. An LDL molecule is composed of about 85% lipid, with cholesterol being its primary lipid constituent. LDLs carry cholesterol to the cells.

HDLs are only composed of 50% lipid. They transport phospholipids and cholesterol. HDL molecules actually transport cholesterol back to the liver.

LIPOPROTEIN LIPASE

Once transported to their destination, fatty acids must be released from the lipoprotein complex. This is done in adipose tissue by an enzyme called *lipoprotein lipase*. The enzyme acts to break triacylglycerol into 2 fatty acid molecules and 2-monoacylglycerol. These molecules then diffuse into fat cells, which reassemble them into triacylglycerol molecules for storage.

Cholesterol Release

Cholesterol carried by LDL is transported to the cells where it can be utilized. Cells have LDL receptors on their surface, which can bind LDL and bring it into the cell by a process of endocytosis. Organelles called lysosomes fuse with the vesicles containing LDL and release lysozymes that hydrolyze the protein-lipid bond. The protein complex is degraded in the cell to its constituent amino acids. At this point, the cholesterol exists in the form of *cholesteryl esters*, which are hydrolyzed into fatty acids and free cholesterol molecules.

β Oxidation of Fatty Acids

Fat is stored as triacylglycerol. To extract energy from the molecule, it must be broken down once again by a lipase enzyme. This enzyme catalyzes the hydrolysis of the triacylglycerol molecule into glycerol and fatty acids. Then, the fatty acids can be utilized to obtain energy in a process that occurs in three steps.

1. Activation
2. Transport to mitochondria
3. Oxidation to acetyl-CoA

We now consider each of these in turn.

ACTIVATION

Activation is a process by which a fatty acid is converted to a *coenzyme a derivative*. This process takes place on the outer surface of the mitochondria. The activation reaction is catalyzed by *acyl-CoA synthetase*. There are three different acyl-CoA synthetase enzymes.

- One to activate acetate, propionate, and butyrate-2 carbon to 4-carbon molecules.
- One that activates medium chain-length fatty acids, which are 4-carbon to 12-carbon molecules.
- Finally, one that activates long-chain fatty acids.

The activation of a fatty acid requires one adenosine triphosphate (ATP) molecule and Coenzyme A (CoASH). *Two phosphate* bonds of ATP are hydrolyzed in the process, leaving behind AMP + PP$_i$ (two atoms of inorganic phosphate). This reaction is illustrated in Fig. 5-2.

TRANSPORT

A transport protein passes acyl-CoA fatty acid derivatives across the mitochondrial membrane to the matrix where they can be oxidized. This is done by *carnitine*, which is an *acyl-group carrier*. Acyl groups are bound to a hydroxyl group of carnitine using an enzyme called *carnitine acyltransferase*. There are actually two carriers, *carrier I* resides on the outer side of the inner mitochondrial membrane, while *carrier II* resides on the inner side of the inner mitochondrial membrane. A

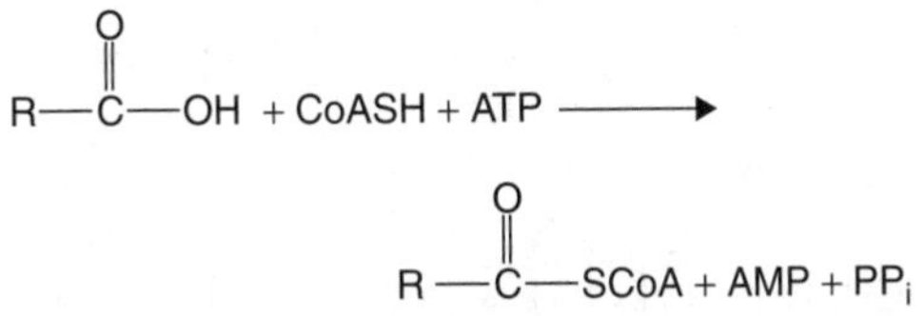

$$R-\overset{\overset{\displaystyle O}{\|}}{C}-OH + CoASH + ATP \longrightarrow$$

$$R-\overset{\overset{\displaystyle O}{\|}}{C}-SCoA + AMP + PP_i$$

Figure 5-2   The activation of a fatty acid.

translocase enzyme moves the acyl-carnitine complex to *acyltransferase II*, which releases the acyl-CoA derivative into the mitochondrial matrix. Free carnitine is transported in the opposite direction, toward the intermembrane space.

OXIDATION

Once inside the mitochondrial matrix, the molecule can be oxidized. This process works by shortening the acyl-CoA fatty acid derivatives two carbons at a time, yielding acetyl-CoA as the end product. This process is called β-oxidation because the bond between C-2 and C-3 of the chain is the β bond. The oxidation process requires four steps.

Step 1: Acyl-CoA → Trans-Δ^2-enoylacyl-CoA (enoyl-CoA)

This reaction reduces an FAD molecule FAD → $FADH_2$. Enoyl-CoA serves as a substrate for water.

Step 2: Enoyl-CoA → 3-hydroxyacyl-CoA

In this step, water is added across the carbon-carbon double bond.

Step 3: 3-Hydroxyacyl-CoA → 3-Keoacyl-CoA

This step reduces NAD+ to NADH.

Step 4: 3-keoacyl-CoA → Acyl-CoA + Acetyl-CoA

This step requires the presence of a CoA molecule. Acetyl-CoA is released from the fatty acid by thiolytic cleavage.

The β-oxidation process continues, removing the first two carbon units of the fatty acid, until no more carbon units are available. For example, palmitoyl-CoA has 16 carbons. Seven cleavages are used which produce 8 molecules of acetyl-CoA which can then be utilized by the mitochondria to produce energy. Under normal conditions, intermediates used in the citric acid cycle are produced by carbohydrates. When glucose is not readily available (such as on a high-protein diet or during vigorous exercise) different mechanism must be used to digest fatty acids. This is done using *ketone bodies*, molecules produced from acetyl-CoA by mitochondria found only in the liver.

Energy Yield

The energy yield from fatty acids is considerably higher than that obtained from glucose. The complete oxidation of one molecule of 16-carbon palmitic acid yields a net of 129 ATP molecules. One molecule of palmitic acid yields 8 molecules of acetyl-CoA from β-oxidation, 7 molecules of flavine adenine dinucleotide and its reduced form (FADH$_2$), and 7 molecules of nicotinamide adenine dinucleotide and its reduced form (NADH). The FADH$_2$ and NADH molecules are oxidized in the electron transport chain. Each molecule of FADH$_2$ yields 2 ATP, thus 14 total ATP are produced from FADH$_2$. Each NADH molecule yields 3 ATP, giving a total of 21 ATP. Each acetyl-CoA yields 1 GTP (guanosine triphosphate) and 12 ATP. With 8 molecules of acetyl-CoA produced from palmitic acid, a total of 8 × 12 = 96 molecules of ATP are produced. A total of 131 molecules of ATP are produced, but there is an energy cost of 2 ATP in the metabolism of palmitic acid, giving the net of 129 ATP molecules. This is about three times the energy liberated from a molecule of glucose.

Summary

The metabolism of lipids takes place using digestive enzymes produced by the pancreas. These enzymes are called lipases and phospholipases. Together with bile salts, they break down fats so that they can be transported to the liver where they are bound together with proteins to form lipoproteins. Lipoproteins can transport lipids through the bloodstream to the cells. Lipids are stored as triacylglyercols, which can be broken down for use by cellular lipases. Lipid metabolism takes place in the mitochondria where large amounts of ATP can be produced from lipids.

Quiz

1. A lipase breaks a triacylglycerol molecule down into
   (a) 2-monoacylglycerol plus one fatty acid.
   (b) 2-monoacylglycerol plus two fatty acids.
   (c) acetyl-glycerol plus two fatty acids.
   (d) 2-monoacylglycerol plus three fatty acids.

2. The lowest density lipoproteins can be best described as
   (a) transporting cholesterol back to the liver.
   (b) transporting cholesterol to cells.
   (c) transporting fat molecules to adipose tissue.
   (d) transporting cholesterol to the small intestine.

3. Chylomicrons are molecules that

   (a) transport CoA lipid derivatives into the mitochondrial matrix.

   (b) are lipoprotein complexes consisting of 95% lipids.

   (c) are lipoprotein complexes consisting of 85% lipids.

   (d) are transport molecules which move fatty acids from the small intestine to the liver.

CHAPTER 6

Protein Structure

Introduction

A *protein* is a large organic molecule constructed of a chain of amino acids. The word protein, derived from the Greek *proteios,* loosely means "holding first place"—an apt name because proteins regulate the life machine. They orchestrate the business of life by controlling multiple bioprocesses including metabolism, cell growth, and neurotransmission. And, although proteins provide structure and can act as energy source, the main reason they are so important is their role as enzymes—enabling chemical reactions that are critical to life. Proteins can do this because they have an ability to move electrons and protons around biosystems.

The sequence of amino acids that form a protein is encoded in our genes, held in the DNA. The collection of proteins produced by a given organism is called the organism's *proteome.* The differences among proteomes of different organisms create different life forms. It is no surprise that researchers throughout the bioscience community are focusing on the huge repertoire of proteins, approximately, 100,000 distinct molecules, produced by the human genome.

To obtain a basic understanding of proteins, you need to know:

- General characteristics of proteins
- The structure of amino acids (the building blocks of proteins)
- How amino acids are linked together
- How the sequence of amino acids on a polypeptide chain results in the three-dimentional structure of the molecule

General Characteristics of Proteins

Biomolecules exhibit a pattern of modular assembly, meaning that they are built from repeating fundamental units. In the case of a protein, the primary structure is a string of *amino acids*. An amino acid molecule is characterized by the presence of an *amine* (NH_2) and a carboxyl group (COOH) on the first (alpha [α]) carbon (Fig. 6-1). The other groups bonded to the α-carbon are represented by the symbol R. The exception is proline, discussed below.

One of the truly remarkable features of proteins is the fact that all of these diverse chemicals are built from only 20 major amino acids. That's really pretty simple, considering the huge diversity of protein function in the life machine. You should be grateful for this, biochemist, since this means *just 20 different molecules* control almost all of our life functions. Not only ours, as human beings, but for *all life forms* on earth. We are all made of the same stuff, from the tiniest mycoplasma to the president of the United States. Note that while there are only 20 major amino acids, there are hundreds of *minor* amino acids. These are amino acids that are either short lived or are simple derivatives of the major amino acids.

To understand why nature can keep things so simple, think of the English alphabet as an analogy. There are only 26 letters in the alphabet that almost any child can learn. Despite this small number, the letters of the alphabet can be arranged in a myriad of different ways to produce wildly different strings of characters or messages.

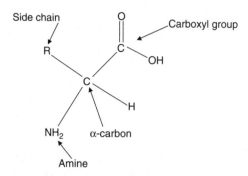

Figure 6-1   Structure of an amino acid.

For example, how many messages can you construct that contain 100 letters? Since there are 26 letter choices for *each* character in the message, there are on the order of

$$26 \times 26 \times \cdots \times 26 \approx 10^{141}$$

possible character messages! This is a very large number indeed.

Similar logic applies to the construction of proteins. There are a very large number of molecules that can be constructed by arranging the 20 major amino acids in chains—about two quadrillion, or 2,000,000,000,000,000 to be exact. This diversity is possible because each point in the peptide chain can hold any one of the 20 amino acids. Also, the length of a protein can be up to several thousand amino acids.

Returning to our analogy with the 26-letter alphabet, the astute reader will note that not all 10^{141} character messages are valid. In fact *most* randomly assembled messages would be completely meaningless. The construction of proteins from amino acids works this way as well. Not all of the two quadrillion combinations of amino acids produce functioning proteins.

To form a polypeptide chain, the carboxyl group of one α-carbon on one amino acid links with the amine group of the α-carbon of the next amino acid. This happens via the removal of an OH from the carboxyl group and an H atom from the amine to form H_2O. The result is a *peptide bond* as shown in Fig. 6-2.

As is typical of the assembly of biomolecules, the formation of a peptide bond requires energy. The result of the initial step of attaching amino acids together through peptide bonds is a long, linear chain of α-carbons in a nice row, called a *polypeptide*. Proteins are initially produced as polypeptides of 300 to 1000 amino acids. Functional proteins may consist of a single, long chain. However, many proteins are composed of two or more chains. Also, some bioactive molecules consist of only about 10 amino acids and are called *oligopeptides*.

Most proteins contain between 50 and 2000 amino acids. When we discuss the configuration of the protein, it will be important to remember that the main chain of the linked α-carbons forms a flat and planar backbone for the protein. The basic structure can bend to an extent—rather like a stiff piece of poster board—but cannot rotate around the peptide bond axis with the freedom of rotation enjoyed by their side chains.

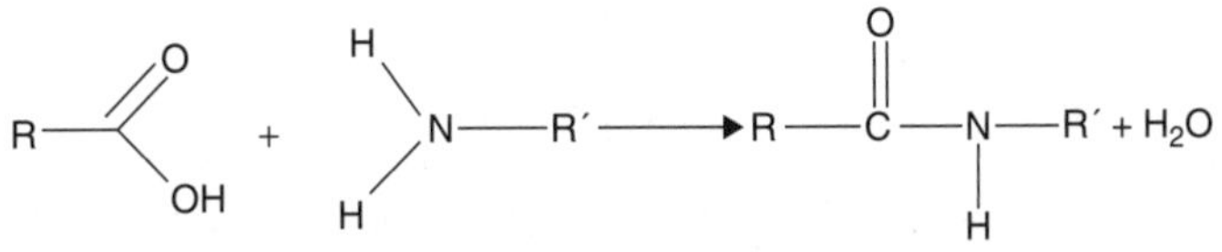

Figure 6-2   Peptide bond.

In addition, the backbone carbon chain is itself rich in hydrogen-bonding potential because each residue has a carbonyl group C=O which is a good hydrogen-bond acceptor (attracts electrons) and an HN group which is a good hydrogen-bond donor (partially gives up electrons). Remember that a hydrogen bond doesn't involve an exchange of electrons, which is simply due to the creation of electromagnetic dipoles within molecules. This capacity to form hydrogen bonds means that, in the backbone of a polypeptide chain, weak attachments can be formed between the amino acids, *tending to bend or twist the chain*.

The α-carbon of an amino acid also bears a single hydrogen and the diverse group symbolized as "R." "R" stands for the side chains. Different amino acids have different R groups attached to their α-carbon. In the primary structure of a protein, the molecules are put together by the ribosomes in a strand, like strings of pearls, and the side chains protrude from the long row of α-carbons. After release from the ribosomes, the R groups interact causing the strand to twist and distort. The final configuration of the protein depends on the order of the amino acids, which determines the interactions between the side groups.

DETAILS OF AMINO ACIDS

Amino acids in solution at neutral pH exist mainly as *zwitterions* rather than as unionized molecules. Zwitterions have both a positive and a negative charge (Fig. 6-3). The amino group carries a positive charge (NH_3^+) and the carboxyl group carries a negative charge (CO_2^-). The presence of both charges means that the molecule can act as both an acid and a base, that is, *amphoteric*.

For reasons unknown to us, all naturally occurring amino acids are in the "L" configuration, meaning simply that, in solution, they rotate a beam of light to the left. Note that if you produce amino acids in a laboratory, you will produce a *racemic mixture* of both L and D configurations. This may be a problem to those of you that plan to use your laboratory-produced molecules in a biological system, because the *unnatural* configuration may not function properly. So biomolecules exhibit *handedness*. (Some bacteria have small amounts of the D-amino acids.)

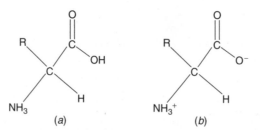

Figure 6-3 The general form of amino acids (*a*) written as a neutral amino acid, and (*b*) written in the zwitterionic form.

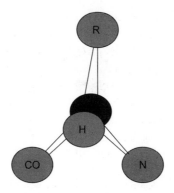

Figure 6-4   Orientation of atoms around the α-carbon of amino acids.

A handy acronym will help you to remember how the groups are arranged around the α-carbon of amino acids (Fig. 6-4). Looking at the molecule from the top with the hydrogen atom sticking toward you, you will observe that the other groups are arranged (from left to right) as C=O, R and NH_2. Reading from left to right and taking the CO, the R and the N, you spell CORN.

We will divide the 20 major amino acids into three groups:

1. Nonpolar side chains

2. Uncharged polar side chains

3. Charged polar side chains

The characteristics of the major amino acids are summarized in Table 6-1. Before we start examining amino acids in detail, it would be wise to review some basic terms in organic chemistry provided in App. A.

In a nutshell, the nonpolar amino acids lack amine, hydroxyl, or sulfhydryl groups in their side chains. The polar side chains have such groups. Whether or not these groups are actually ionized at a physiological pH depends on the pKa for the dissociation of the protein member of the group.

Nonpolar Side Chains

With the exception of glycine, proteins with nonpolar side chains avoid the watery environment of the cell. These consist of amino acids whose side chains are long chains of methyl and methylene groups and are hydrophobic. These molecules do not ionize and, because the carbon and the hydrogen share the electron equally, these molecules do not exhibit electromagnetic poles that create hydrogen bonding. There are no ionizable protons. These are *glycine, valine, alanine, leucine, isoleucine, proline, phenylalanine,* and *tryptophan.* Their side groups contain just carbon and hydrogen, as methyl or methylene groups. These hydrophobic groups tend to cluster

Table 6-1 Summary of Characteristics of Major Amino Acids

Name	Abbreviation(s)	Unique Features	Structure
Nonpolar, hydrophobic side chains—found in protein interiors			
Glycine	Gly, G	• Smallest • R is a single proton • Flexible • Neuroinhibitor • Role in biosynthesis of many compounds, such as purines	^+H_3N—C—COO$^-$ (H, H)
Alanine	Ala, A	• Methane • α-Keto homologue is pyruvate • Role in nitrogen transport from tissues to the liver	CH_3 ; ^+H_3N—C—COO$^-$; H
Valine	Val, V	• Butyl group • Branched chain amino acid found in high concentration in muscles	CH_2 ; H—C—CH_3 ; ^+H_3N—C—COO$^-$; H
Leucine	Leu, L	• Branched chain amino acid found in high concentration in muscles	CH_3 ; H—C—CH_3 ; H—C—H ; ^+H_3N—C—COO$^-$; H
Isoleucine	Ile, I	• Isomer of leucine • Branched chain amino acid found in high concentration in muscles	CH_3 ; CH_2 ; H_3C—C—H ; ^+H_3N—C—COO$^-$; H

Table 6-1   Summary of Characteristics of Major Amino Acids (*Continued*)

Name	Abbreviation(s)	Unique Features	Structure
Nonpolar, hydrophobic side chains—found in protein interiors			
Methionine	Met, M	• One of two amino acids containing sulfur • Contains a sulfur as an ester • Sulfur easily oxidized • As s-adenosyl-L-methionine (SAM), a methyl donor in many bioreactions	
Proline	Pro, P	• α-Carbon forms a ring containing primary amine • Inflexible • Forms kinks in secondary structures	
Phenylalanine	Phe, F	• Alanine plus a phenyl • Converted to tyrosine, which is, in turn, converted to L-dopa • Interferes with the production of serotonin	
Tryptophan	Trp, W	• Bulky, aromatic side chains • Indole group • Precursor for serotonin and niacin	

(*Continued*)

Table 6-1 Summary of Characteristics of Major Amino Acids (*Continued*)

Name	Abbreviation(s)	Unique Features	Structure
Uncharged polar side chains—metabolically active and located on the exterior of proteins			
Serine	Ser, S	• Hydroxyl group • Found at the active site of enzymes • Aids in glycoprotein formation	OH | CH_2 | ^+H_3N—C—COO^- | H
Threonine	Thr, T	• Methyl and hydroxyl group • Found at the active site of enzymes • Aids in glycoprotein formation	OH | H_3C—C—H | ^+H_3N—C—COO^- | H
Asparagine	Asn, N	• Methyl group and carboxyl group with a highly polar uncharged amine that readily forms hydrogen bonds • Found at the ends of alpha helices and beta sheets • Aids in formation of glycoproteins • Input to urea cycle • α-Keto homologue is oxaloacetate	O NH_2 $\backslash\!/$ C | CH_2 | ^+H_3N—C—COO^- | H
Glutamine	Gln, Q	• Two methyl groups and carboxyl group with a highly polar uncharged amine • Forms isopeptide linkages • Central role as nitrogen donor in synthesis of nonessential amino acids • Provides nitrogen transport to the liver	O NH_2 $\backslash\!/$ C | CH_2 | CH_2 | ^+H_3N—C—COO^- | H

Table 6-1   Summary of Characteristics of Major Amino Acids (*Continued*)

Name	Abbreviation(s)	Unique Features	Structure
Uncharged polar side chains—metabolically active and located on the exterior of proteins			
Tyrosine	Tyr, Y	• Similar to phenylalanine but with polar hydroxyl group on phenyl ring • Important metabolically because ionization altered by micro pH changes	(structure)
Charged polar side chains—reactive			
Cysteine	Cys, C	• Sulfhydryl (thiol) group • Forms disulfide bridges • Found at the active site of enzymes • Binds iron	(structure)
Lysine	Lys, K	• Long aliphatic chain terminating in an amine • Nucleophilic • Forms ionic bonds	(structure)
Arginine	Arg, R	• Long aliphatic chain containing an amine and terminating in two amines • Nucleophilic • Forms ionic bonds • Generated in the urea cycle	(structure)

(*Continued*)

Table 6-1 Summary of Characteristics of Major Amino Acids (*Continued*)

Name	Abbreviation(s)	Unique Features	Structure
Charged polar side chains—reactive			
Histidine	His, H	• Methyl • Imidazole group • Ionic bonds found at the active site of enzymes • Crucial in the structure of hemoglobin	
Aspartate or aspartic acid	Asp, D	• Donates an amine to become oxaloacetate • Active in proteolytic enzymes	
Glutamate or glutamic acid	Glu, E	• Central role as nitrogen donor in synthesis of nonessential amino acids • Provides nitrogen transport to the liver	

together in the interior of the molecule and thereby to shield themselves against the aqueous environment. Glycine, however, can be found in the interior or the exterior of a protein.

Two members of this group have a great effect on the final configuration of a protein. Glycine, the simplest amino acid, has an R group that is simply a proton. Glycine's small size enables it to fit into tight spots. Remember this when we talk about collagen.

In contrast, proline has a side chain of methylenes covalently bonded to the nitrogen atom of the backbone. This forms a five-membered pyrrolidine ring that rigidly holds the amino acid in a single conformation. The presence of proline can cause kinks in an otherwise smoothly folded peptide chain. Proline initiates some of the structures regularly found in proteins such as the α-helix. Also, no hydrogen bonding is possible with the α-carbon because no amide hydrogen is present.

Uncharged Polar Side Chains

The second group consists of amino acids whose side chains are hydrophilic because they have either hydroxyl (OH), sulfhydryl (SH), or amine (NH_2) groups. But, even though they contain ionizable groups, these groups are minimally ionized at the neutral pH of the cell. The group contains *serine, threonine, asparagine, glutamine, tyrosine,* and *cysteine.* Remember that these side groups are not charged but exhibit electromagnetic dipoles and can participate in hydrogen bonds. These dipoles are created because hydrogen and oxygen, hydrogen and sulfur, and hydrogen and nitrogen do not share electrons equally. Because they are uncharged, these amino acids also tend to occur on the inner side of the protein molecule. However, unlike the nonpolar amino acids, they stabilize protein structure and enable folding by participating in interior hydrogen bonds. Also, the ability to exchange protons or electrons in cellular microenvironments of varying pH makes these amino acids potentially active in the fundamental business of regulating bioprocesses.

Cysteine is very important in influencing the structure of a protein. In cysteine-rich proteins, two cysteine residues are frequently covalently linked to each other (through oxidation of the SH groups) to produce a disulfide bridge. The resulting unit of two molecules of cysteine is called *cystine* (Fig. 6-5). Disulfide bonds

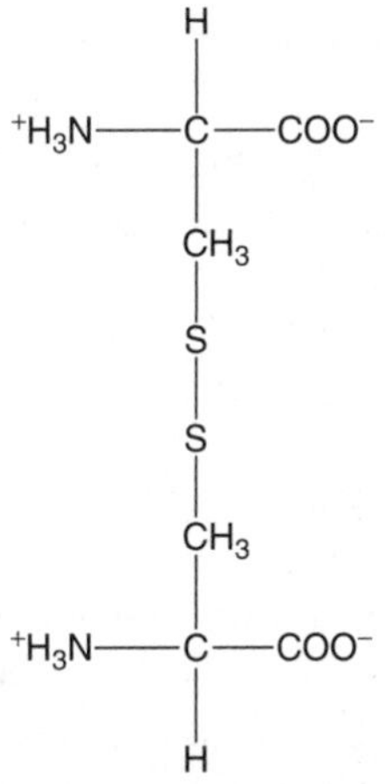

$$
\begin{array}{c}
H \\
| \\
^+H_3N\!\!-\!\!-\!\!C\!\!-\!\!-\!\!COO^- \\
| \\
CH_3 \\
| \\
S \\
| \\
S \\
| \\
CH_3 \\
| \\
^+H_3N\!\!-\!\!-\!\!C\!\!-\!\!-\!\!COO^- \\
| \\
H
\end{array}
$$

Figure 6-5   Formation of cystine.

crosslink proteins and stabilize the protein structure. Proteins are held together primarily by relatively weak hydrogen bonds, so disulfide bonds can have a very large effect on the stability of the protein structure. Extracellular proteins often have disulfide bonds whereas intracellular proteins usually lack them.

Sulfhydryl groups are very important in other ways. The sulfhydryl group commonly occurs at the active site of enzymes. Also, the sulfhydryl groups can form a complex with iron (Fe), forming Fe-S proteins, or *ferredoxins*.

Serine and threonine contain carboxylic acids (COOH). The hydroxyl groups can accept or donate a proton giving them an important role in the biosystem machinery. They are important in a number of active functions, as well as participating in structural H-bonding. Serine is located at active site of a number of enzymes, where the hydroxyl group participates in hydrolytic reactions. Serine and threonine also aid in the binding of nonamino acid molecules to proteins. For example, the hydroxyl group at the surface of the protein can be used to build oxygen bridges and to bind sugars, making glycoproteins.

The hydroxyl group of tyrosine is poorly ionized at neutral pH because the ring structure tends to hold on to the electron. However, ionization increases at more basic pHs which can occur in a microenvironment of a biochemical reaction. Tyrosine is found at the active site of some enzymes. The ionization of the side chain can be increased by manipulation of the local pH to create basic conditions.

Unlike serine and threonine, asparagine, and glutamine are amides rather than carboxylic acids. Their side chains do not ionize and are not very reactive. However, the amide group is very polar and participates in hydrogen bonds as both hydrogen bond donor and acceptor.

Glutamine participates in linkages between protein chains in a special covalent linkage, called an isopeptide bond. Isopeptide bonds are similar to peptide bonds except the amino group participating in the bond is not the α-amino group. This link leads to proteins that are tightly bound, like keratin molecules in hair and in formation of a fibrin blood clot.

Asparagine is used to bind sugars to proteins to form glycoproteins. Interestingly, unlike normal cells, some cancer cells cannot make asparagines and must acquire this amino acid from the diet. It is thought that if cancer cells can be denied a source of asparagine, this could offer an innovative cancer treatment option.

Charged Polar Side Chains

The amino acids with charged side chains at neutral pH are *lysine, arginine, histidine, aspartic acid,* and *glutamic acid. Lysine* and *arginine* are basic because of the presence of amines. Lysine is an active *nucleophile* (donate a pair of electrons to a chemical bond) and participates in a variety of reactions. Arginine contains a guanidine (i.e., has three nitrogen atoms) and is also strongly basic.

Histidine has an *imidazole* ring in the side chain (a ring structure containing nitrogen). This makes it nucleophilic (electron rich) and basic like other amines. In the nonprotonated form of imidazole, the nitrogen-hydrogen is a hydrogen bond donor while the carbon-bonded nitrogen is a hydrogen bond acceptor. The imidazole ring renders histidine very versatile. Histidine is found in the active sites of enzymes because the imidazole ring can bind and release protons in the course of enzymatic reactions. Aspartic and glutamic acid are important in many biosynthetic reactions. The conjugate base of glutamic acid, glutamate, plays a crucial role in nitrogen metabolism as a nitrogen donor to form nonessential amino acids and as a nitrogen acceptor in transporting nitrogen to the kidneys.

In summary, the 20 major amino acids provide great versatility within the protein structure. The carbon backbone itself is rich in structures that form hydrogen bonds. The hydrophobic side chains of some amino acids cause portions of the molecule to behave like a lipid and seek the inside of the molecule structure. Others are ionized and found on the outside of the protein molecule where they can interact with their environment. Many of the side chains form hydrogen bonds within a protein. Two can form strong covalent bonds either within one polypeptide chain or between polypeptide chains, one involving sulfide atoms (cysteine) and one utilizing a carboxyl and amide groups of the side chains in a bond, like the peptide bond (glutamine). Those amino acids possessing surface hydroxyl groups or highly basic nitrogen groups tend to be found at the active side of enzymes because of their ability to accept or donate electrons.

FORMATION OF PROTEINS FROM AMINO ACIDS

Proteins are constructed on the ribosomes, where the genetic code is implemented via strands of RNA. The way this is done is discussed in detail in a later chapter. For our purposes, suffice it to say that the sequence of amino acids is dictated by the genetic code and read out by the ribosomes. As you go through this discussion, look at it like an entrepreneur—an entrepreneur who is interested in the production of proteins.

The key to the proper function of many proteins is their *shape*. The shape is formed as the protein folds. Proteins with exactly the same amino acids, but arranged in different order, can be shaped differently and hence perform a different function. And, as we will see below, identical proteins may or may not share the same function, depending on how they are folded. If folded improperly, proteins may fail to function or may accumulate into insoluble aggregates known as *inclusion bodies*.

The shape of a protein develops in stages. The structure of the original protein as it is created on the ribosome is known as the *primary structure*. The primary structure is a long string of amino acids forming *polypeptide chains*.

When complete, the primary polypeptide chain looks rather like a lengthy caterpillar—only it is flat. Extruding from this creature are numerous protrusions. Each segment of this caterpillar bares quite different protrusions. The protrusions are endowed with forces that affect the movement of the caterpillar. One protrusion is very tiny and can fit into a small space, should the creature curl around tightly. One is unusually bulky and tends to spread things out. One forms a loop upon itself and creates a knot. Some seek protection from their environment. These tend to be found inside the creature when it curls up. Others exhibit an attraction to one another (through hydrogen bonds) and induce the creature to curl. Some form attachments either between protrusions (because they are ionized through presence of hydroxyl or amine groups). One protrusion bonds with a like protrusion, creating an unusually strong force to hold the creature in a given configuration.

Protein structures generally are described at four levels: primary, secondary, tertiary, and quaternary. A *primary structure* is simply the two-dimentional linear sequence of amino acids in the peptide chain. See Fig. 6-6.

The secondary structure is formed as the protein begins to twist in accordance with chemical forces within the primary chain. The hydrogen in the amine groups and the oxygen in the carboxyl groups of the backbone form hydrogen bonds that stabilize the secondary structures.

Secondary structure commonly takes one of two forms. One form is a left-handed helix that is developed as the molecule twines around itself. This form is called an

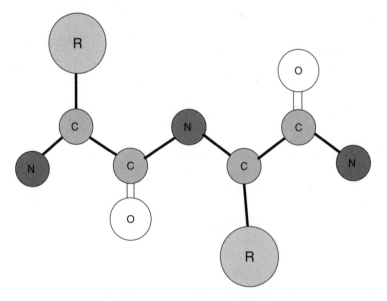

Figure 6-6   Protein primary structure.

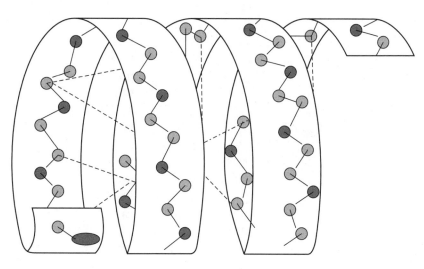

Figure 6-7 α-Helix.

α-helix (Fig. 6-7). As is usual with biological terminology the alpha designation has nothing whatsoever to do with this configuration, but is part of the name because this was the first configuration discovered. In the α-helix, stabilizing hydrogen bonds formed between amine and carboxyl groups on the carbon backbone are tight. The side chains protrude from the exterior of the helix.

Globular proteins typically contain the α-helix form. As the name implies, globular proteins are globelike in form and do not have structural strength. Hormones are globular proteins. Also, some structural proteins also consist of α-helices. For example, *keratin* is a family of structural *fibrous proteins* that include α-keratin. Hair and other α-keratins consist of coiled single protein strands, which are then further coiled into superhelical structures. These structures can stretch out and return to the original configuration. The supercoils contain a high percentage of cysteine residues that form stabilizing cross-links. The sulfur from the cysteine accounts for the distinct odor of burning hair. Extensive cysteine cross-links between α-coils form the hard structures of hooves and nails.

The other secondary form is a crimped shape called a *β-pleated sheet*. A β-pleated sheet looks a little like a piece of tin (Fig. 6-8). The initial polypeptide chains form links between carboxyl and amine groups on adjacent chains. β-Sheets can consist of peptide chains that run in opposite directions (antiparallel) or in the same direction (parallel) (Fig. 6-9). Either way, the hydrogen of amine groups on one chain form hydrogen bonds with the oxygen of carboxyl groups on another chain. β-Sheets contain a high percentage of the smallest amino acids, glycine, and alanine. The lack of bulky side chains usually allows tight linkage of adjacent peptide strands. β-Sheets are usually found in structural proteins, such as silk and the *β-keratin proteins* found in claws, beaks, and shells of turtles.

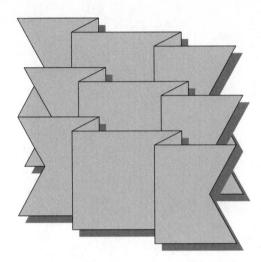

Figure 6-8   β-Pleated sheet.

The more intricate folding begins with the tertiary structure as the protein folds back on itself (Fig. 6-10). As the helix or sheet bends back on itself in a precise order, bonds are formed between the side chains, including hydrogen and ionic bonds.

A quaternary structure may be formed from the association of two or more peptide strands. The formation of disulfide bonds is important in the quaternary structure of many proteins. The quaternary structure of many proteins includes both α-helixes and β-sheets.

If a protein is disassembled such that the structure disintegrates, the protein is said to be *denatured*. Examples of factors that can denature proteins include heat and changes in pH. Note that even though secondary and tertiary structures are gone, the primary structure remains, with the peptide bonds intact. When denatured,

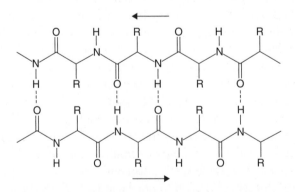

Figure 6-9   Antiparallel β-sheet.

Figure 6-10   Tertiary structure.

the basic polypeptide strand forms loose loops called a *random coil*. If conditions are right, this random coil may spontaneously reassume the appropriate shape through self-assembly.

Many proteins are built from discernable units that can assemble and disassemble independently of the rest of the protein. These building units are known as *domains*. Some of these domains are common building units that recur across protein families. For example, many enzymes contain a substrate binding site called the α/β-barrel. The α/β-barrel is also called the TIM barrel after triose phosphate isomerase, which was the first such structure to be described. The barrel is formed through linked α-helixes and β-sheets that curve around in such a way as to form a fat barrel, or donut, shape. The β-strands form the inner wall of the donut and the α-helices form the outer wall of the donut. The active site is on the outside of the "barrel." The center of the barrel is stuffed with hydrophobic amino acid side chains. See Fig. 6-11.

Some proteins contain unique structures. An example is collagen, the fibrous protein found in skin, cartilage, bone, and other connective tissue and the most common human body protein. Collagen, which composes more than a third of the body's total protein content, is important enough to spend a little time in description.

Collagen is formed from subunits called *tropocollagen,* a rod made up of three polypeptide strands (Fig. 6-12). These three left-handed helices are twisted together into a right-handed triple helix stabilized by numerous hydrogen bonds. The result is a versatile molecule that, depending on the assembly, can provide varying degrees of flexibility to body structures. If the triple helix structure is disrupted, as by heat, the result is, literally, *gelatin.*

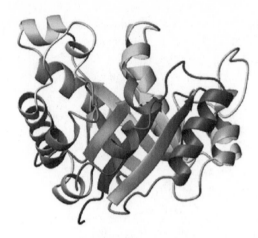

Figure 6-11   Example of quaternary structure—TIM barrel.
(*Courtesy of Wikipedia commons.*)

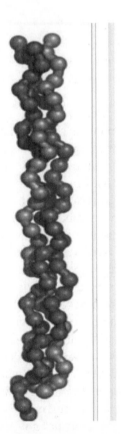

Figure 6-12   Tropocollogen. (*Courtesy of Wikipedia commons.*)

You may remember the discussion on amino acids and the fact that glycine's small shape allows it to fit into tight spots and the proline's ring structure tends to constrain mobility. Collagen consists of approximately a third glycine and about 10% proline. The glycine allows for tight coupling of adjacent strands and the proline provides rigidity at strategic points within a strand. Collagen also contains *hydroxyproline*, which is derived from proline and *hydroxylysine*, which is derived from lysine. Synthesis of both of these derivative amino acids requires vitamin C. This role explains why lack of vitamin C causes a painful joint condition called *scurvy*.

Protein Folding and Misfolding

The proper folding of the proteins engages an eloquent cellular system that is crucial to the functioning of the proteins and the survival of the cell.

Self-Assembly

Christian Anfinson, winner of the 1972 Nobel Prize in chemistry, was the first researcher to observe a protein fold spontaneously into its tertiary structure in the laboratory. The secondary and tertiary structure of the protein, ribonuclease A, was destroyed using the denaturing agents—urea and mercaptoethanol. When these denaturing agents were removed, Anfinson and his colleagues observed that the protein refolded spontaneously.

The conclusion from this research was that the forces necessary to mold the protein into the correct shape are *embedded in the structure of the protein* and the cellular environment is not necessary to cause the protein to fold correctly. In other words, the final configuration is achieved by *self-assembly*. Furthermore, since a large number of folded configurations are possible but a protein finds the correct configuration rapidly, the folding process must follow a specific, predetermined pathway. Subsequent research has indicated that the primary force leading to specific protein configurations is the push to shield nonpolar side chains of certain amino acids. Portions of the polypeptide chain containing such amino acids seek the inside of the folded structure. Formation of electrostatic and hydrogen bonds then stabilize the final structure.

One problem is that this very intricate folding process needs to occur in the complex chemical soup of the cell cytoplasm. Eukaryotic cells manage to isolate protein production somewhat by confining it largely to the matrix of the endoplasmic reticulum. Prokaryotic cells attempt something of the same thing by conducting protein synthesis in the periplasm just underneath the cell membrane. The location of these protein factories close to the cell membrane is fortuitous because many proteins are destined to be excreted or will become part of the cell membrane as receptors.

In some cases, proper folding does not just happen in accordance with the chemical forces within the cell. It is actually regulated by a general system known

as *folding moderators*. The moderators include *folding chaperonins* and *folding catalysts* that accelerate rate-limiting steps.

There are basically four functions of molecules that act as chaperonins: (1) enable proper folding, (2) hold partially folded molecules until the system has the capacity to finish the folding activity, (3)usher proteins into and out of the cell membrane, and (4) disassemble misfolded proteins.

Enabling Proper Folding

As the name implies, folding chaperonins help the protein to fold properly. Many such chaperonins fall under the general name of *heat shock proteins* (HSP). HSPs are so named because they are found in greater quantities in cells that are experiencing stress of some sort, such as temperature or pH extremes. One of the HSPs that have been studied in detail is the GRO/EL system, first described in *Escherichia coli*. The GRO/EL system forms a hollow structure that provides a nifty sequestered microenvironment within which proteins may safely fold (Fig. 6-13). The system

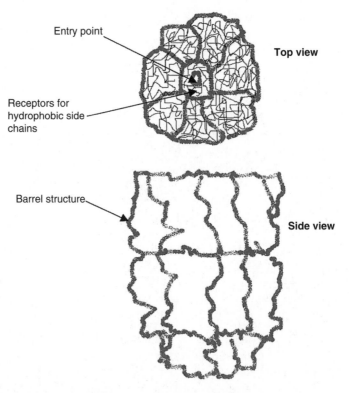

Figure 6-13   GRO/EL schematic.

projects hydrophobic units into the cell environment. These units are sought and bounded by exposed hydrophobic units on proteins. Remember that these units are supposed to be sequestered on the interior of a properly folded protein. If they are sticking out, the protein probably needs to be refolded. Once a protein is bound, configurational changes within the system bring the protein into the interior, which forms a cave around the protein and provides an optimal environment for disaggregation of misfolded proteins, folding and refolding. The increase in GRO/EL under condition of stress reflects that fact that the stress increases the probability of misfolding. GRO/EL and other like systems will sequester new proteins until the cell environment is more suitable for their function.

Sometimes proteins are designed to be refolded. The protein may require folding in one way for transport to the cell membrane only to be refolded to perform its function. One of the reasons that proteins sometimes are folded and then refolded is the need to transport them through the aqueous environment of the cytoplasm and insert them into or extrude them through the cell membrane. In bacteria, the periplasm contains chaperonins who assist the folding and membrane insertion of outer membrane proteins. In eukaryotic cells, these chaperonins are found primarily in the lumen of the endoplasmic reticulum and in the Golgi apparatus.

The successful exportation of proteins depends on: (1) a *signal* or *targeting sequence* on the protein that marks it for insertion into the membrane and (2) a secretory modulator that recognizes this sequence. Many of the proteins destined to be exported are produced as a "preprotein" and contain a unique series of amino acids that exist solely to mark the protein for secretion or insertion into the plasma membrane. These serve as the targeting sequences.

The pathway a protein takes may be dependent on specific *secretory chaperonins*. The protein to be extruded is transported to the export site and is bound by a secretory chaperonin targeted to the special signal sequence of the protein. The signal sequences can mark the protein for excretion via a specific secretory pathway. The chaperonin may drive itself into the cell membrane, carrying the export protein along. In the process, the preprotein may be clipped at a proteolytic cleavage site to provide the protein its final conformation.

In eukaryotic cells, most of the posttranslational modification of proteins destined for exportation or for insertion into the cell membrane occurs in the lumen of the endoplasmic reticulum (ER) or in the Golgi apparatus, which is devoted to the sequestering and manipulation of proteins. Proteins are brought into these structures by a system similar to the one described above whereby the protein bears a signal sequence recognized by other molecules that will bind the protein and bring it into the ER.

To fold correctly, some molecules must undergo a series of steps. The intermediate steps require insertion of small molecules that act as fasteners or covers that are later removed. These small chaperonins function like devices in an assembly line that are attached to the product, such as a clip or a temporary holding device, during manufacture, but must be removed before the product is completed. Another fitting analogy would be a surgical clamp that is removed before the surgical incision is closed. Chaperonins include small sugars that attach to the protein as it folds. A typical function of these molecules is to cover hydrophobic surfaces. The continued exposure of hydrophobic areas of a protein is an indication that the protein has been misfolded.

The typical target for these mini-chaperonins is a short stretch of hydrophobic amino acids, a very common structure in protein polymers. Because the binding site is nonspecific and common, bioengineers may be able to select various nonspecific chaperonins that will enable proper protein folding in environments where it would not otherwise occur. In some biosystems that mimic disease conditions related to misfolded proteins, the misfolding has been corrected or prevented by the addition of small biomolecules selected because of their ability to bind to proteins. Nonspecific chaperonins may provide treatment for diseases caused by protein misfolding, such as Alzheimer.

Summary

Proteins orchestrate cell functions. The repertoire of proteins contained in each living organism is what distinguishes one life form from another. Proteins play many important roles in the body, acting as energy sources and providing structure to skin, cartilage, and bone. However, the main reason that proteins are so important is their role as enzymes—moving the business of life along by enabling chemical reactions that are critical to life. Proteins can do this because of their ability to move electrons and protons around biosystems.

Proteins are composed of amino acids. Amino acids are characterized by an amine group (NH_2) and a carboxyl group (COOH) on the first or α-carbon. Curiously, amino acids at physiological pH have both a positive (NH_3^+) and a negative (COO^-) charge. This means that the amino acid can act as both an acid and a base. Proteins are formed by strings of amino acids that align like strands of pearls as the protein emerges from the ribosome.

Amino acids differ in their side chains, identified by the symbol "R" on the α-carbon. There are three classes of amino acids, depending upon the ability to bear a charge—that is, the ability to accept or release a proton. The first class has no ability to exchange protons with the environment because it consists of amino acids with long hydrophobic, aliphatic side chains. By definition, such side groups are lipids because they avoid the watery environment of the cell and tend to cluster together. The latter tendency leads to the folding of a polypeptide chain such that the hydrophobic side chains are sequestered inside of the protein. The second class consists of amino acids with amine, hydroxyl, or sulfhydryl groups. These groups can bear a charge, but they are not ionized at physiological pH. This class of amino acids is useful in enzyme function because the ionization state can be changed by changing the pH of the cell microenvironment and thereby changing the ability to accept or donate a proton. The third class of amino acids is charged at physiological pH. These amino acids tend to be very bioactive and are found at active sites of enzymes and in functions such as neurotransmission.

Several of the amino acids are especially important in determining the configuration of a protein. These are glycine, whose small size enables tight configurations, and proline, which has the opposite role of providing rigidity at strategic locations. Cysteine is notable because of the ability to form strong covalent bonds within and among polypeptide strands.

After amino acids are strung into a polypeptide chain, the chain undergoes configuration changes to form the protein secondary structure. Two configurations are common, an α-helix and a β-sheet. Most proteins contain a combination of these two forms. Other important secondary structures include that of collagen, the most common protein in the human body. Collagen consists of intertwined rods that exist in subunits of three, called tropocollagen. When collagen is denatured, the rods unwind and the result is gelatin.

The final form of most proteins requires further configuration changes, creating the tertiary structure, which can consist of linked β-sheets and α-helixes, and the quaternary structure. The latter is often composed of multiple polypeptide chains that exist in domains, or modules that tend to recur as common configurational themes within different types of proteins.

In 1972, Christian Anfinson observed that under the proper conditions in the laboratory, denatured ribonuclease A spontaneously reformed the correct structure. This observation led to the conclusion that proteins self-assemble into the correct secondary, tertiary and quaternary structure. Since this time however, we have learned that the cell often assists proteins in folding, especially when cell conditions are less than optimal. The molecules that assist folding and refolding of misfolded proteins are called chaperonins. Many chaperonins have been observed to increase in abundance under conditions of thermal stress and hence are called heat shock proteins (HSP). Some HSP systems form caves around newly formed proteins and provide an optimal environment for folding. They also may encase and disassemble misfolded proteins, giving them another chance for proper folding. These systems sequester proteins from adverse conditions during times of environmental stress. Recently, small nonspecific molecules, including some sugars, have been observed to bind to misfolded proteins and to enable proper refolding. There may therefore be an opportunity to intervene in disease conditions related to misfolded proteins by introducing these small enabling chaperonins into diseased cells.

QUIZ

1. The backbone of a protein forms the basis of the secondary structure stabilized by

   (a) sulfhydryl covalent bonds.

   (b) ionic bonds.

   (c) hydrogen bonds between the side chains.

   (d) hydrogen bonds between groups on the α-carbon.

   (e) All of the above.

2. In a folded protein, one would expect

   (a) hydrophilic groups on the outside where they interact with their environment.

   (b) hydrophobic groups on the inside where they stabilize the protein by ionic bonds.

   (c) both hydrophilic and hydrophobic groups on the inside forming hydrogen bonds.

   (d) hydrophilic groups on the inside where they stabilize the protein by ionic bonds.

3. Proline is likely to be

   (a) ionized at physiological pH.

   (b) found in areas where protein structure exhibits rigidity.

   (c) found at active sites of enzymes.

   (d) important in formation of hydrogen bonds in the protein backbone.

   (e) Both (a) and (c).

4. Histidine is

   (a) nonpolar.

   (b) electron rich.

   (c) a kink in a polypeptide chain.

   (d) found on the inside of protein structures.

5. α-Helices and β-sheets are different because

   (a) α-helices are formed within a single polypeptide chain and β-sheets are formed between polypeptide chains.

   (b) α-helices are secondary structures and β-sheets are tertiary.

(c) only β-sheets are found in keratin.

(d) β-sheets are held together by cysteine linkages, whereas, α-helices are formed by hydrogen bounds.

6. Gelatin is

(a) denatured collagen.

(b) an α-helix.

(c) common domain.

(d) tropocollagen.

7. Glycine is

(a) found in rigid portions of the protein.

(b) found at active sites of enzymes.

(c) found in tight curves.

(d) polar but nonionized at physiological pH.

8. The following statement is not true:

(a) Vitamin C is required for the synthesis of amino acids found in collagen.

(b) Many proteins self-assemble into the correct shape.

(c) Proteins in general require chaperonins for proper folding.

(d) Secretory proteins contain a unique sequence recognized by special chaperonins.

(e) Heat shock proteins increase in times of cellular stress.

9. The following is not true about heat-shock proteins:

(a) Are found in greater abundance in cells under thermal stress.

(b) Are harvested by heat-shocking the cells and collecting the precipitate.

(c) Are important in protein folding.

(d) Are important in protecting cells from the impact of stressful environments.

10. The following statement is not true:

(a) Hydrogen bonds are much weaker than the bond that forms cystine.

(b) pH in the cell microenvironment may be different than the rest of the cell.

(c) Many secretory proteins require refolding prior to insertion into the cell membrane.

(d) There are only 20 distinct amino acids.

(e) A denatured protein forms a random coil.

CHAPTER 7

Protein Assembly and Disassembly

Introduction

The regulation of protein production, recycling, and destruction is important for many reasons. First, proteins determine everything else, so producing a protein sets other reactions in motion. Protein production takes enormous energy. Therefore, the production must be efficient and directed toward the immediate needs of the cell. The need for efficiency drives the mechanisms for conserving nitrogen and recycling amino acids. Protein destruction generates a potentially toxic by-product, ammonia, associated with the nitrogen. Systems for regulating the recycling and elimination of nitrogen are crucial for cell survival.

We will review the mechanism for interpreting the genetic information on the DNA and thereby producing the proteome from amino acids available in the cell in Chap. 14. This chapter will address the cell mechanisms to acquire nitrogen from the atmosphere, to produce specific amino acids, and to recycle and eliminate proteins.

You will need to know:

- How nitrogen is acquired from the atmosphere
- Mechanisms for and regulation of amino acid production
- How proteins are recycled and eliminated
- How protein turnover is regulated

A Few Words about Oxidation-Reduction Potential

As we plow through the machinery to make and unmake proteins, you will observe that protons and electrons are inserted at strategic points. Where do these come from? There are several types of biomolecules that serve as reservoirs for both protons and electrons. Protons can be held in reserve as part of an amino acid. Recall in the discussion of amino acids, that some structures create the ability of the molecule to accept or donate a proton. Most notably, amino acids with carboxyl-containing side chains function in proton exchange, explaining why they are instrumental in active sites of enzymes.

Also, cells contain molecular capacitors that can store electrons. Such molecules are known as redox proteins because they can mediate electron exchange between other molecules. Foremost of these tiny capacitors is a class of molecules called *ferredoxins* (Fig. 7-1). The ferredoxins contain iron-sulfur assemblies that can store and release electrons by altering the oxidation state of the iron molecules in these assemblies.

Acquisition of Nitrogen from the Environment

Proteins are possible only because biosystems have evolved to release nitrogen from the very stable nitrogen gas state, and insert the nitrogen into biomolecules. The process to do this is called *nitrogen fixation*. The triple bond in molecular

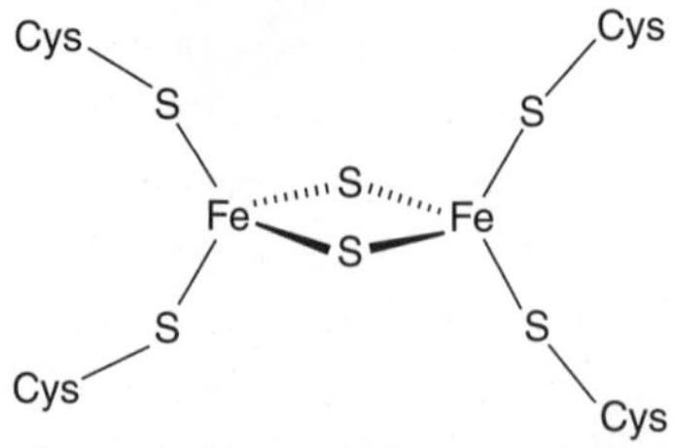

Figure 7-1   Ferredoxin molecule.

nitrogen (N_2) is one of the strongest in nature. Nitrogen fixation is a general term for the process by which stable triple bonds that secure atmospheric nitrogen (N_2) are broken, and nitrogen atoms are incorporated into nitrogen compounds (such as ammonia and nitrate). In the latter, reduced forms, nitrogen is available to biosystems.

We depend on specific types of prokaryotes to perform this crucial operation on atmospheric nitrogen. Microorganisms that fix nitrogen are called *diazotrophs*. They are endowed with the operative class of enzymes, *nitrogenases*. If you want to worry about something, worry about the sudden disappearance of diazotrophs from the face of the earth. Biological fixation of nitrogen is responsible for half of the total removal of atmospheric nitrogen—the rest is accomplished by industry and by lightning.

Biological nitrogen fixation is accomplished by the following reaction:

$$N_2 + 8H^+ + 8e^- + 16\,ATP \rightarrow 2NH_3 + H_2 + 16ADP + 16\,P_i$$

There are several observations you should make about this reaction. First, the process requires a great deal of energy input, 16 molecules of adenosine triphosphate (ATP). The the conversion of ATP to adenosine diphosphate (ADP) and P_i releases 3.4 kJ/mol of free energy. To produce the ATP, 10 g of glucose is required for every gram of ammonia produced. Secondly, toxic ammonia is produced. The ammonia is ionized to ammonium (NH_4^-). You will learn in this chapter how the nitrogenase-generated ammonia is removed from solution by assimilation into glutamate through the glutamine synthetase/glutamate synthase pathway.

The eight electrons required for nitrogen fixation come from ferredoxin and are transferred through an enzyme called *nitrogenase reductase*. The ferredoxin, in turn, receives electrons from a *hydrogenase* enzyme, found in many nitrogen-fixing bacteria. The hydrogenase enzyme harvests electrons from H_2. Additional sources of electrons may include nicotinamide adenine dinucleotide (NADH), flavoproteins, and nicotinamide adenine dinucleotide phosphate (NADPH). See Fig. 7-2 for a summary of this process.

The nitrogenase enzymes (also known as *molybdenum-iron protein*) are very susceptible to destruction by oxygen (another thing to worry about). Diazotrophs have evolved means to reduce the oxygen pressure in their environment. They thrive by living in confined anaerobic or partially anaerobic conditions (underground) and they engage special proteins, such as *leghaemoglobin,* to bind oxygen.

Some nitrogen-fixing bacteria and blue-green algae live in symbiosis or in association with green plants, including peas, beans, clover, and soy. Most of these are nodule bacteria (e.g., *Rhizobium*) of the family Leguminosae. Their host plants are called *legumes.*

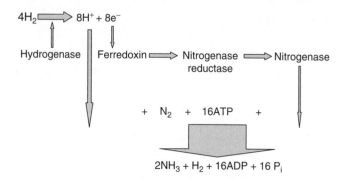

Figure 7-2   Nitrogen fixation.

There are also soil bacteria that can harvest nitrogen from decaying organic matter. These are called *nitrifying bacteria*. The nitrogen is released from biomolecules as either nitrate or nitrite and is eventually returned to the atmosphere as N_2. This process (one of the many circles of life) is called the *nitrogen cycle*. Note that most microorganisms (by the way, not you) are capable of reducing $NO_3^- + NO_2^-$ to NH_3^-.

The genetic basis of nitrogen fixation has been well characterized. Genetic engineering offers the possibility that the ability to fix nitrogen can be endowed upon other organisms, possibly even green plants. Economic incentives to do this are enormous. Plants would be, in essence, self-fertilizing. Predictably, progress has been slow. It is said predictably because the process is complex. Not only must the nitrogenase enzymes be provided, but also ferredoxin for electron storage, the nitrogenase reductase to move the electrons from the ferredoxin, and hydrogenase to provide electrons to the ferredoxin. Technical challenges include ensuring the gene is expressed in eukaryotic systems, providing oxygen-free compartments within the host organism for the function of the nitrogenase and providing sufficient energy to drive the process. Bioengineering thrusts to increase the nitrogen-releasing efficiency of the soil bacteria have been more successful.

Essential versus Nonessential Amino Acids

The human body can produce the following amino acids:

- Alanine

- Asparagine

- Aspartic acid

- Cysteine
- Glutamine
- Glutamic acid/glutamate
- Glycine
- Proline
- Serine
- Tyrosine

These 10 are called nonessential amino acids because it is not necessary that we include them in our diet. By contrast, the human body cannot produce the following amino acids:

- Arginine
- Histidine
- Methionine
- Threonine
- Valine
- Isoleucine
- Lysine
- Phenylalanine
- Tryptophan
- Leucine

Consequently, these 10 must be included in the human diet.

It is important to note that the actual need of specific amino acids depends on the age of the organism and on stress levels. For example, only human infants typically need arginine and histidine. Some amino acids considered nonessential should be supplemented under certain conditions. For example, asparagine tends to be consumed in the urea cycle faster that it is produced, so it may become a dietary essential. Cysteine and tyrosine, both nonessential, are formed from methionine and phenylalanine, both essential amino acids. If the latter two are low, cysteine and tyrosine can become essential. There is evidence that, although glutamate is considered nonessential, supplements of glutamate may be necessary to fulfill the substantial need for this crucial amino acid. If so, the controversial food supplement monosodium glutamate (MSG) may have gotten a bad rap.

It is truly essential that your body receive the necessary complement of amino acids. If there are deficiencies in amino acids, the body has only one source for these—the body proteins, especially muscle. Faced with amino acid deficiencies,

muscle tissue, including such disconcerting sources as heart muscle, will be degraded regardless of the calorie content and other nutritional features of the diet. It is literally possible to starve to death even when provided with ample calories.

INCORPORATION OF AMMONIUM INTO NONESSENTIAL AMINO ACIDS

The Formation of Glutamine by Incorporation of Ammonia

Reduced nitrogen enters the human body as dietary free amino acids, protein, and the ammonia produced by intestinal tract bacteria. Ammonium becomes part of a biomolecule by the following reaction:

$$\text{Glutamate} + \text{NH}_4^+ + \text{ATP} = \text{glutamine} + \text{ADP} + \text{P}_i + \text{H}^+$$

This reaction is catalyzed by the enzyme *glutamine synthetase*. The carboxyl group (COO$^-$) at the tail end of glutamate is exchanged for an amide (NH$_2$), creating glutamine. Let's look at the molecular structure of glutamate and glutamine (Fig. 7-3).

Transfer of Amines to α-Keto Acids

The strategy to build most amino acids is to transfer an amine group, usually from glutamine, to an *α-keto acid*. α-Keto acids are a group of molecules whose carbon skeletons correspond to those of the amino acids except that one carbon bears a double bounded oxygen (called a ketone) and they lack the amine group of an amino acid α-carbon. The molecule is reduced by eliminating the ketone body (the oxygen) and adding an amine (the nitrogen). See Fig. 7-4.

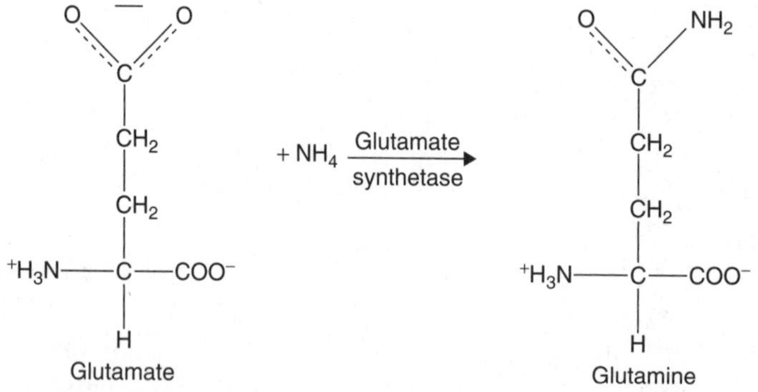

Figure 7-3 Formation of glutamine from glutamate and ammonium.

$$\overset{\text{NADPH} \quad \text{NAD}^+ + \text{H}^+}{\curvearrowright}$$

$$\underset{\substack{\text{O} \\ \alpha\text{-Keto acid}}}{\text{R}-\overset{\|}{\text{C}}-\text{COOH}} + \text{NH}_3 \longrightarrow \underset{\substack{\text{NH}_2 \\ \text{Amino acid}}}{\text{R}-\overset{\overset{\text{H}}{|}}{\text{C}}-\text{COOH}} + \text{H}_2\text{C}$$

Figure 7-4   Conversion of an α-keto acid to a corresponding amino acid.

The glutamine donating the amine is thereby transformed into glutamate. These amine-transfer reactions are called *transamination* and are catalyzed by a group of enzymes called *transaminases,* also called *aminotransferases.* Aminotransferases all require *pyridoxal phosphate* (PLP), a vitamin B$_6$ derivative, as a cofactor.

The highest volume of transamination reactions involves the creation of alanine and aspartate, corresponding to the fact that the precursors of these amino acids are the common metabolites, pyruvate and oxaloacetate. Clinically, the presence of alanine aminotransferase in the serum is interpreted to mean heart or liver damage.

This is a good point for a little summary. Nitrogen is reduced from atmospheric nitrogen to a biologically available form by nitrogen-fixing bacteria. This reduced form, ammonia, is incorporated into biomolecules by glutamine synthetase which replaces a carboxyl group on glutamate with an amine, converting it into glutamine. The glutamine is the primary source of amino groups that create all of the other nonessential amino acids by transfer of an amine group to the corresponding α-keto acids, regenerating glutamate. The transfer of the amine group is accomplished by aminotransferases.

The Role of Glutamate in Managing Nitrogen Inventory

The carbon skeletons of many amino acids can feed into *gluconeogenesis,* or synthesis of glucose, when the amino acid is removed. See Chap. 10 for a description of gluconeogenesis. The α-keto acids of glutamate, alanine, and asparagate, especially, are part of the gluconeogenesis cycle. Therefore, if a cell, such as a muscle, needs to harvest energy, these amino acids can be converted to glucose in the liver with the nitrogen eliminated as urea in the kidney. Alternatively, if structural or active proteins are needed, nitrogen can be added to the corresponding α-keto acid, creating the necessary molecule. The most important amino acid in the regulation of nitrogen is—you guessed it—glutamate.

Glutamate is reduced to its α-keto acid by the enzyme *glutamate dehydrogenase.* This enzyme can act in reverse and serves a key role in linking catabolic and metabolic pathways. The reaction catalyzed by glutamate dehydrogenase is:

$$\text{NH}_4^+ + \alpha\text{-ketoglutarate} + \text{NADPH} + \underset{\text{ATP}}{\rightleftharpoons} \text{glutamate} + \text{NADP}^+ + \text{ADP} + \text{P}_i$$

The glutamate dehydrogenase is controlled by relative availability of ATP and ADP. ATP inhibits the breakdown of glutamate; whereas ADP is an activator. Therefore, breakdown of glutamate is active when cellular energy is needed, as indicated by buildup of ADP. Protein is created when cellular energy levels are high, as indicated by buildup of ATP.

Nitrogen balance is an important index of homeostasis. An organism in equilibrium is neither losing nor acquiring nitrogen. Nonetheless, nitrogen must be cycled, depending on the needs of the organism. Some nitrogen must be recycled to produce new protein, and some nitrogen must be eliminated in the urine as protein is broken down. I think you can appreciate how important the ability to shift an amine group between glutamine and glutamate is in maintaining nitrogen equilibrium. Consider the fact that an average human requires 28 g of protein per day, but approximately 400 g is broken down, recycled, or eliminated each day. Nitrogen regulation is a responsive and closely controlled system.

Also the glutamine-glutamate system provides a way for the body to transport a toxic substance, ammonia, to a site where it can be reused, that is, the liver, or a site where it can be eliminated, that is, the kidney. For transport, the ammonia is in a nontoxic form, attached to glutamate, which becomes glutamine (Fig. 7-5). Glutamine is the most common amino acid in the serum.

Although the glutamine-glutamate cycle is foremost in transport, recycling, and elimination of nitrogen, alanine serves a related role. Alanine is ideal because its α-keto acid is pyruvate, the input carbohydrate for gluconeogenesis. In the liver, *alanine transaminase* transfers ammonia from alanine to α-ketoglutarate and regenerates glutamate and pyruvate. This process is referred to as the *glucose-alanine cycle.* Using the glucose-alanine cycle, skeletal muscle can eliminate nitrogen while replenishing energy supply.

$$
\begin{array}{ccc}
\begin{array}{c}
O \\
\parallel \\
C-O^- \\
| \\
C=O \\
| \\
CH_2 \\
| \\
CH_2 \\
| \\
C-O^- \\
\parallel \\
O
\end{array}
&
\begin{array}{c}
+\,NH_3 \quad \xleftarrow{\qquad} \\
\xrightarrow{\qquad} \\
\text{Glutamate} \\
\text{dehydrogenase}
\end{array}
&
\begin{array}{c}
O \\
\parallel \\
C-O^- \\
| \\
HC-NH_3^+ \\
| \\
CH_2 \\
| \\
CH_2 \\
| \\
C-O^- \\
\parallel \\
O
\end{array}
\\
\text{α-Ketoglutarate} & & \text{Glutamate}
\end{array}
$$

Figure 7-5   Action of glutamate dehydrogenase.

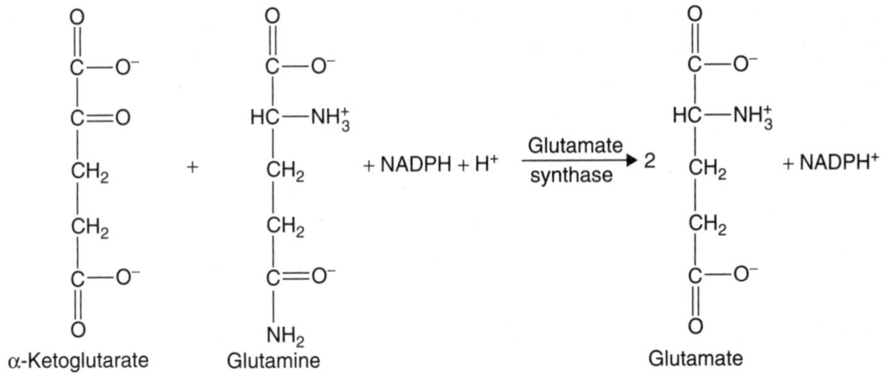

Figure drawing (chemical structures):

α-Ketoglutarate + Glutamine + NADPH + H⁺ →(Glutamate synthase)→ 2 Glutamate + NADP⁺

$$\alpha\text{-Ketoglutarate} + \text{Glutamine} + NADPH + H^+ \xrightarrow{\text{Glutamate synthase}} 2\ \text{Glutamate} + NADP^+$$

Figure 7-6 Action of glutamate synthase.

Although the body is very sensitive to ammonium concentration, some ammonia is present in the serum in equilibrium with ammonium ion. The liver acts to regulate the ammonia-ammonium concentration to nontoxic levels.

Bacteria possesses another enzyme called *glutamate synthase*. This enzyme uses glutamine as a nitrogen source to produce glutamate from α-ketoglutarate. See Fig. 7-6.

Note that this reaction sequence uses both NADPH and ATP to incorporate one molecule of ammonium—the extra energy expenditure is required to assimilate nitrogen under limiting conditions. Bacteria, which possess two pathways of nitrogen assimilation, regulate the process so that the energy intensive glutamate synthase reaction is active only when ammonium levels are low. When ammonium levels are high, glutamate dehydrogenase is active and glutamine synthetase and glutamate synthase are not.

Table 7-1 should help sort out which enzyme does what.

Biosynthesis of Nonessential Amino Acids

Alanine, asparagine, aspartate, glutamate, and glutamine are synthesized from a carbonaceous skeleton provided by pyruvate, oxaloacetate, or α-ketoglutarate. These amino acids are created by a single transamination reaction.

Glutamate is the precursor of proline and arginine. The arginine is formed from ornithine in the urea cycle and ornithine is derived from glutamate. In these reactions, glutamate provides the carbon backbone, and glutamine supplies the amine. See Fig. 7-7 for a depiction of the synthesis of proline from glutamate.

Serine, and glycine are derived from 3-phosphoglycerate with the amine donated from glutamine (Fig. 7-8). Cysteine is synthesized from serine and homocysteine, a breakdown product of methionine (Fig. 7-9).

Table 7-1   Nitrogen Metabolic Enzymes

Enzyme	Source	Enzyme Function
Hydrogenase	Bacteria—diazotrophs	Harvest electrons from hydrogen and pass to ferredoxin
Nitrogen reductase	Bacteria—diazotrophs	Collects electrons from ferredoxin and pass to nitrogenase
Nitrogenase	Bacteria—diazotrophs	Produce reduced nitrogen (ammonia) from atmospheric nitrogen.
Glutamine synthetase	All cells	Adds ammonium ion to glutamate, producing glutamine.
Glutamine dehydrogenase	All cells	Remove amine from glutamate, producing α-keto glutamate and vice versa
Glutamine synthase	Bacteria	Converts α-ketoglutarate to glutamate using glutamine as a nitrogen source.
Aminotransferases	All cells	Transfer amine from glutamine to various α-keto acids, creating essential amino acids.

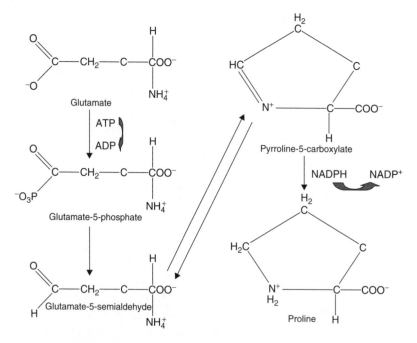

Figure 7-7   Biosynthesis of proline from glutamate.

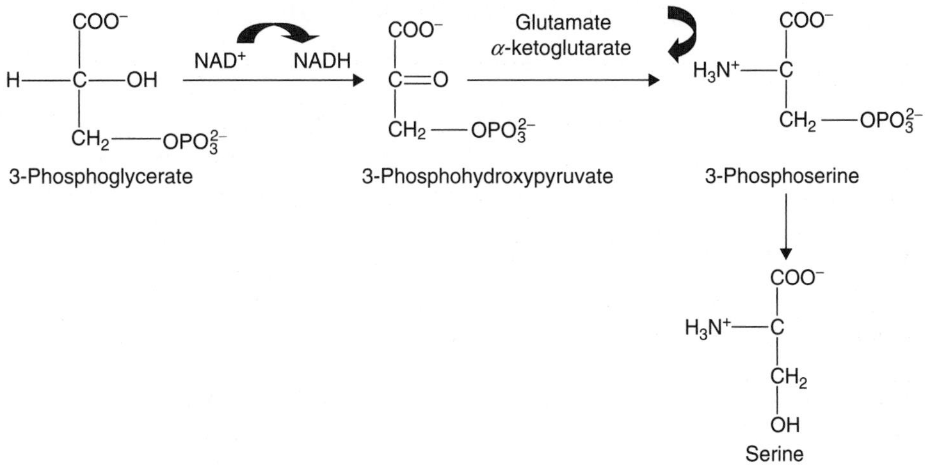

$$
\begin{array}{ccccc}
\underset{\text{3-Phosphoglycerate}}{\begin{array}{c} COO^- \\ | \\ H\!-\!\!\!-C\!-\!\!\!-OH \\ | \\ CH_2\!-\!\!\!-OPO_3^{2-} \end{array}}
& \xrightarrow[NAD^+ \quad NADH]{}
& \underset{\text{3-Phosphohydroxypyruvate}}{\begin{array}{c} COO^- \\ | \\ C\!\!=\!\!O \\ | \\ CH_2\!-\!\!\!-OPO_3^{2-} \end{array}}
& \xrightarrow[\text{α-ketoglutarate}]{\text{Glutamate}}
& \underset{\text{3-Phosphoserine}}{\begin{array}{c} COO^- \\ | \\ H_3N^+\!-\!\!\!-C \\ | \\ CH_2\!-\!\!\!-OPO_3^{2-} \end{array}}
\end{array}
$$

$$
\underset{\text{Serine}}{\begin{array}{c} COO^- \\ | \\ H_3N^+\!-\!\!\!-C \\ | \\ CH_2 \\ | \\ OH \end{array}}
$$

Figure 7-8 Biosynthesis of serine from 3-phosphoglycerate with amine donation from glutamate.

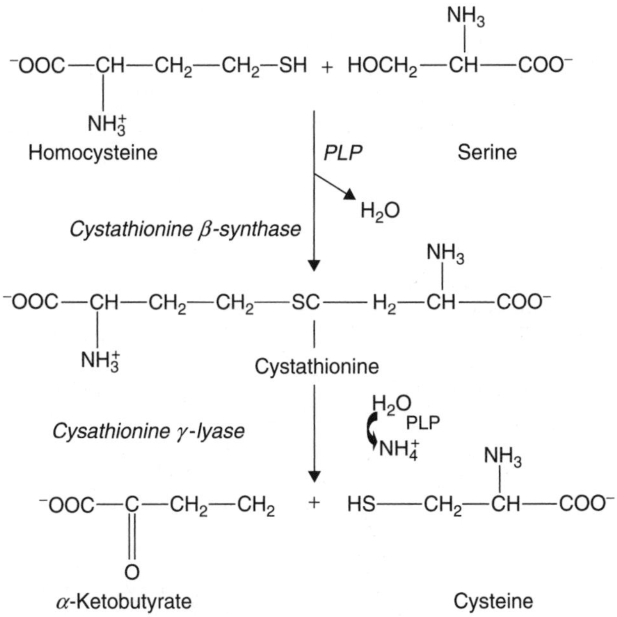

$$
\underset{\text{Homocysteine}}{{}^-OOC\!-\!\underset{\underset{NH_3^+}{|}}{CH}\!-\!CH_2\!-\!CH_2\!-\!SH} \;+\; \underset{\text{Serine}}{HOCH_2\!-\!\overset{\overset{NH_3}{|}}{CH}\!-\!COO^-}
$$

Cystathionine β-synthase   PLP → H_2O

$$
\underset{\text{Cystathionine}}{{}^-OOC\!-\!\underset{\underset{NH_3^+}{|}}{CH}\!-\!CH_2\!-\!CH_2\!-\!SC\!-\!H_2\!-\!\overset{\overset{NH_3}{|}}{CH}\!-\!COO^-}
$$

Cysathionine γ-lyase H_2O PLP NH_4^+

$$
\underset{\text{α-Ketobutyrate}}{{}^-OOC\!-\!\underset{\underset{O}{\|}}{C}\!-\!CH_2\!-\!CH_2} \;+\; \underset{\text{Cysteine}}{HS\!-\!CH_2\!-\!\overset{\overset{NH_3}{|}}{CH}\!-\!COO^-}
$$

Figure 7-9 Biosynthesis of cysteine from serine.

Regulation of cysteine production is important because high levels of the precursor homocysteine correlate with cardiovascular disease.

Tyrosine is made by hydroxylation of the essential amino acid phenylalanine. A genetic deficiency in the metabolism of phenylalanine causes a condition called *phenylketonuria* (PKU). In this disorder, the enzyme that converts phenylalanine to tyrosine (phenylalanine hydroxylase) is deficient. As a result, phenylalanine and its breakdown products, accumulate in the blood and body tissues, resulting in severe mental retardation. If this condition is diagnosed in a timely fashion in newborns, the symptoms can be prevented by dietary restrictions on phenylalanine.

As you can see, the word "nonessential" is somewhat unfortunate, because some of these amino acids have precursors that are "essential" (Table 7-2). If your diet is deficient in the precursor, then the product also becomes deficient and thereby essential.

Biosynthesis of Essential Amino Acids

Essential amino acids are manufactured by microorganisms and plants. Biosynthesis starts with common metabolites but the enzymes necessary to carry out the metabolic pathway were lost early in animal evolution. Of the 10 essential amino acids, three—lysine, methionine, and threonine—are synthesized from aspartate. The metabolic paths tend to be more complex than those available in animals for synthesis of nonessential amino acids. For example, lysine is derived from aspartate through a multistep process involving donation of an amine from glutamate.

Table 7-2 Summary of Biosynthesis of Essential Amino Acids

Amino Acid	Precursor	Process
Alanine	Pyruvate	Single transamination
Asparagine	α-Oxaloacetate	Single transamination
Aspartic acid	α-Oxaloacetate	Single transamination
Glutamine	α-Ketoglutarate	Single transamination
Glutamic acid	α-Ketoglutarate	Single transamination
Arginine	Glutamine	Through ornithine intermediate in urea cycle
Proline	Glutamine	Pyrrole-5 carboxylate intermediate
Serine	3-Phosphoglycerate	Transamination from glutamine
Glycine	3-Phosphoglycerate	Transamination from glutamine
Cysteine	Serine and homocysteine	Homocysteine produced by breakdown of methionine

Even though plants are the source of essential amino acids, it is difficult to receive a full complement of what you need by eating plants. For example, cereals, the staple of the diet in many developing countries, are deficient in lysine. A diet of cereal-type plants must be supplemented by pod seed, such as peas and beans, to provide all of the essential amino acids. A diet which includes meat is more likely to fulfill our nutritional needs for amino acids.

Role of Folic Acid and Vitamin B$_{12}$

Folic acid is a form of the water-soluble vitamin B$_9$. The name of folic acid is derived from the Latin name for leaf. Folic acid participates in the reduction/oxidation reactions that involve single carbons and is especially important in amino acid metabolism. The folic acid is present to accept methyl groups, usually from serine, and to donate these groups to an appropriate acceptor.

Folic acid includes the compound para-aminobezoic acid. Sulfanomide resembles para-aminobenzoic acid and, when present in bacteria, inhibits their ability to make folic acid. Because we ingest ready-made folic acid in our diet, we are not affected by the sulfanomide antibiotics.

Vitamin B$_{12}$ is required in the metabolism of many single carbon entities. This vitamin consists of a cobalt metal bound to a porphyrin ring. It has broad functions in metabolism of complex organisms. Vitamin B$_{12}$ is also required by bacteria to activate methyl groups that then participate in methionine biosynthesis.

CELLULAR METABOLISM OF PROTEINS

The cellular system that controls protein metabolism includes the capability to destroy proteins and recycle the amino acids within the cell. Proteolytic enzymes within the cell are associated with structures called *proteasomes*. Proteasomes are described in detail in Chap. 2. They are large, dome-shaped protein complexes found in eukaryotic cells, archaea, and some bacteria. Misfolded proteins attach to the dome surface of the proteasomes and then are folded into an interior cavity, where they are snipped into small pieces. This process prevents the accumulation of abnormal polypeptides within the cell and, because the amino acids can be recycled, conserves the energy invested in producing amino acids. Most proteins that are degraded have undergone multiple cycles of folding and misfolding.

In eukaryotic systems, the destruction of proteins is preceded by the adherence of markers. These markers are so ubiquitous in eukaryotic cells that they are called *ubiquitin*. The tagging reaction is catalyzed by enzymes called *ubiquitin ligases*. Once a protein is tagged with a single ubiquitin molecule, this is a signal to other ligases to attach additional ubiquitin molecules. The result is a *polyubiquitin chain* that is bound by the proteasome, resulting in degradation of the tagged protein.

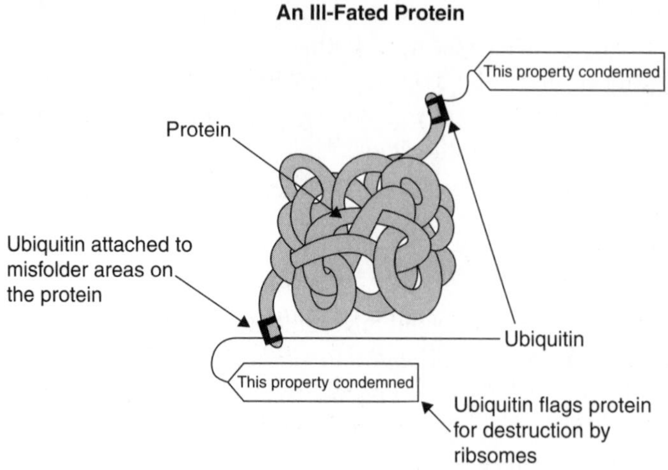

An Ill-Fated Protein

Figure 7-10 A protein marked for destruction.

Once an ubiquitin molecule binds to the unfortunate protein, many others attach and attract the destroyer proteosome. See Fig. 7-10.

Protein Digestion

Dietary proteins must be broken down before they can be absorbed from the gastrointestinal tract. A class of extracellular proteolytic enzymes called *proteases* break down proteins into short peptide chains in a process called *proteolysis*. Proteases tend to be nonspecific and attack many different peptide bonds between the protein amino acids.

Proteases are secreted into the intestinal tract in an inactive form called *zymogen*. This strategy protects the pancreases and the cells of the digestive tract from being digested by their own products. Once in the intestinal tract, intestinal enzymes cleave off portions of the zymogen, creating the active proteolytic enzyme. The most important of these proteases are trypsin, chymotrypsin, carboxy peptidase, secreted by the pancreas; and pepsin, elastase, and amino peptidase, secreted by gastric cells. Other gastric enzymes control the pH of the intestinal tract. For example, the enzyme gastrin causes the release of HCl into the stomach, reducing the pH to 2.0 and thereby helping to breakdown food products and eliminates pathogens ingested with the food. In the small intestine, secretin is produced by gastric cells when food is present and causes the release of bicarbonate. The bicarbonate reduces the pH and allows the effective functioning of proteases and other digestive enzymes.

The small peptides produced can then be transported into the cell where they are further degraded into amino acids. The amino acids can then be transformed into

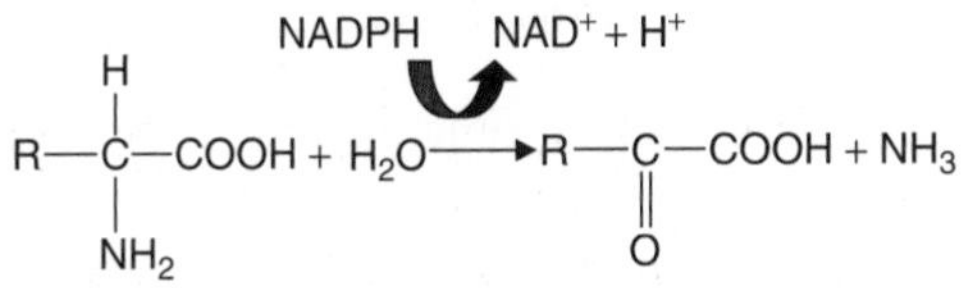

$$\text{R}-\underset{\underset{\text{NH}_2}{|}}{\overset{\overset{\text{H}}{|}}{\text{C}}}-\text{COOH} + \text{H}_2\text{O} \xrightarrow{\qquad\text{NADPH}\quad\text{NAD}^+ + \text{H}^+\qquad} \text{R}-\underset{\underset{\text{O}}{\|}}{\text{C}}-\text{COOH} + \text{NH}_3$$

Figure 7-11   Creation of an α-ketoacid from an amino acid by oxidative transamination.

carbohydrates by the process of *oxidative deamination,* whereby an amine is removed. During oxidative deamination, an amino acid is converted into the corresponding keto acid by removing the amine functional group and replacing it with a ketone group. See Fig. 7-11. The ammonia produced eventually goes into the urea cycle and the keto acid can participate in gluconeogenesis or in the *tricarboxylic acid (TCA) cycle.*

The deamination reaction is shown in Fig. 7-11.

Look familiar? This is the reaction on glutamate catalyzed by glutamate dehydrogenase. Glutamate dehydrogenase also enables the reverse reaction, hence the name "dehydrogenase."

Deamination leaves a carbon/hydrogen skeleton. Deamination of several of the amino acids produces molecules that may be immediately metabolized because they are intermediates in the TCA cycle and other major metabolic pathways (Table 7-3). Examples are 2-oxoglutarate, produced from glutamate, oxaloacetate, produced from aspartate and pyruvate produced from alanine and serine. Deamination of other amino acids leads to products that are handled by specific catabolic pathways before entry into glycolysis or the TCA cycles. As you know, nonessential amino acids can be recreated, primarily by transamination of carbon skeletons.

In the human body, oxidative deamination takes place in the liver. The substrates usually are either alanine, aspartate, or glutamate. Most deamination, however, occurs on glutamate, the end product of most transamination reaction.

Table 7-3   Products Created by Deamination of Amino Acids

Amino Acid	Product
Glutamate	2-Oxoglutarate
Aspartate	Oxaloacetate
Alanine	Pyruvate
Serine	Pyruvate
Valine	2-Oxoisovalerate
Leucine	2-Oxoisocaproate

Urea Cycle

The flow of nitrogen in and out of an organism must be regulated to ensure that enough nitrogen is available to build essential biomolecules but that the toxic by-product, ammonia, does not accumulate. Humans are sufficiently sensitive to the toxicity of ammonia that the serum concentration is allowed to fluctuate very little.

Single cell organisms and aquatic organisms can simply allow the ammonia to diffuse outside of their boundaries. However, multicellular organisms have to work a little harder. In most higher organisms, nitrogen is excreted as urea, a small water soluble molecule containing a single carbon atom and two amides. Urea is also known as *carbamide*. Birds and reptiles secrete an insoluble form of nitrogen as uric acid. The secretion of insoluble uric acid in place of urea allows these animals to conserve water because no water is involved to dissolve the secretory product. See Fig. 7-12.

Urea is secreted by means of the *urea cycle*. Like all metabolic processes, the process consists of a transfer of ions between intermediates resulting in a final product or output. The intermediate molecules are recycled into the metabolic process. In this cycle, amino groups donated by ammonia and L-aspartate are converted to urea, while L-ornithine, L-citrulline, L-argininosuccinate, and L-arginine act as intermediates (Fig. 7-13). The strategy is to transfer two amino groups to a carbon (other than the α-carbon) of an amino acid with no such groups and then split off the carbon bearing these groups as urea, regenerating the original. Here is the cycle in a nutshell:

The input to the process is a very simple molecule, *carbomyl phosphate*. The amide is scavenged from free ammonia or from the glutamine-glutamate cycle and added to a bicarbonate ion. The carbomyl phosphate presents the nitrogen to the urea cycle in an activated form (Fig. 7-14).

$$NH_4^+ + HCO_3^- + 2ATP = \text{Carbomyl phosphate} + 2ADP + P_i$$

The enzyme *ornithine transcarbamoylase* splits off the phosphate and allows carbomyl phosphate to bond with *ornithine*, forming *citrulline* (Fig. 7-15).

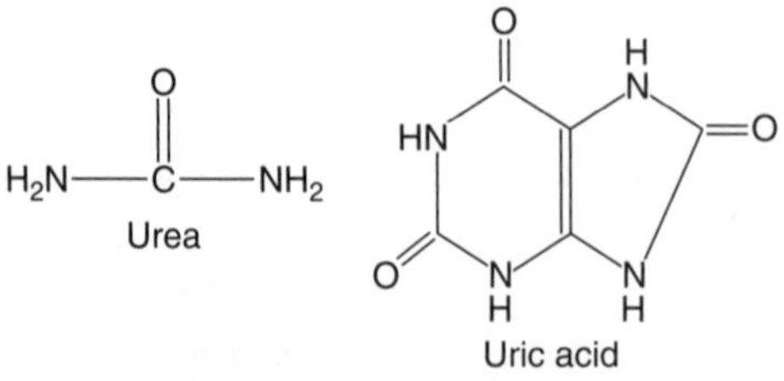

Figure 7-12 Urea and uric acid.

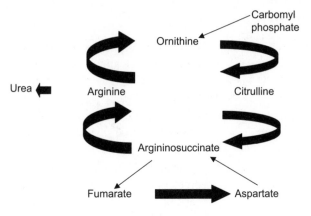

Figure 7-13   Urea cycle.

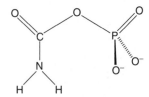

Figure 7-14   Carbomyl phosphate.

Figure 7-15   Formation of citrulline from carbamoyl phosphate and ornithine.

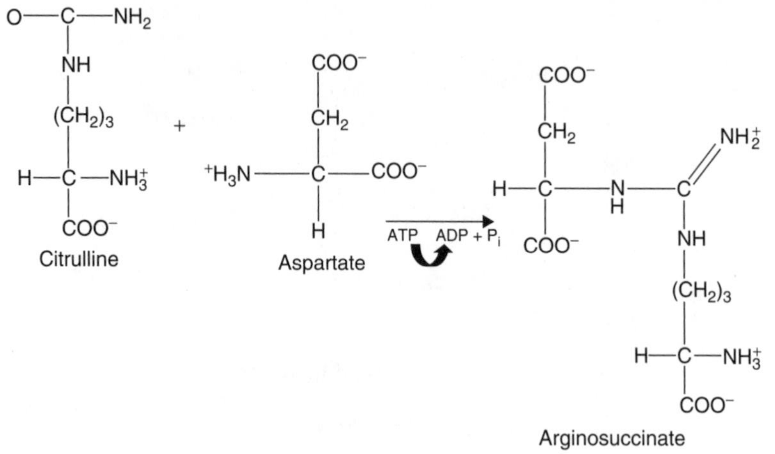

Figure 7-16 Formation of arginosuccinate from citrulline and aspartate.

In a reaction that requires energy from ATP, the citrulline and aspartate are joined via a Schiff base, to form *arginosuccinate* (Fig. 7-16). This reaction accomplishes the goal of attaching two amides to a non–α-carbon. ATP is cleaved in this step.

Now the molecule is split at a site away from the Schiff base, forming arginine and splitting off the 4-carbon hydrocarbon, fumarate. *Arginosuccinase* catalyzes this reaction (Fig. 7-17).

The terminal carbon of arginine is cleaved by *arginase* to form urea and ornithine, and the cycle begins again (Fig. 7-18).

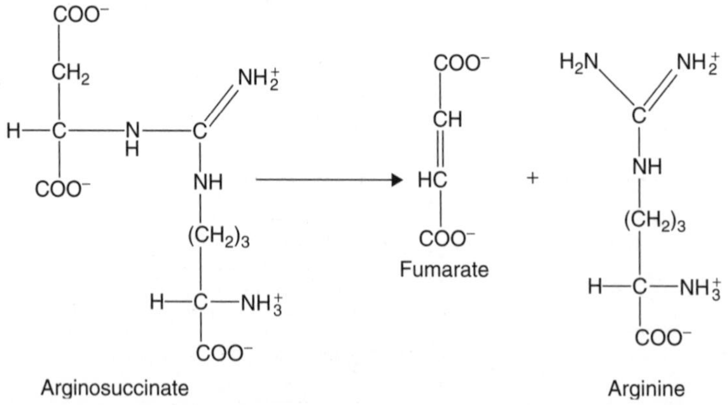

Figure 7-17 Splitting of arginosuccinate to form arginine and fumarate.

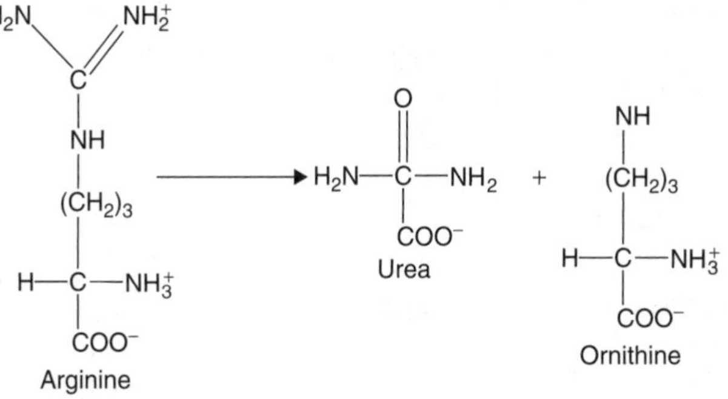

Figure 7-18 Formation of urea from arginine.

Happily, aspartate can be reformed from fumarate, a reaction requiring the old familiar glutamate dehydrogenase. See Table 7-4 for principle enzymes of the urea cycle.

Regulation of Protein Metabolism

Regulation of protein metabolism is a critical process for maintenance of healthful conditions. The concentration of a given protein may dictate the speed and magnitude of a critical process. To ensure that protein concentrations can be finally tuned to environmental conditions, many proteins have short half-lives in the cell. Such proteins must be replaced at a level responsive to the needs of the cell. Bacteria have developed especially sophisticated mechanisms for the regulation of both catabolic and anabolic pathways because their environment is subject to rapid change.

Protein synthesis is regulated at the level of DNA transcription. The initiation of readout of many genes is dependent upon local conditions. Readout may be inhibited

Table 7-4 Principle Enzymes of the Urea Cycle

Enzyme	Function
Ornithine transcarbamoylase	Strips phosphate from carbomyl phosphate and combines skeleton with ornithine to form citrulline.
Arginosuccinase	Forms arginate and fumarate from arginosuccinate.
Arginase	Cleaves the terminal carbon of arginine to form urea and ornithine.
Glutamate dehydrogenase	Reforms aspartate from fumarate.

by the product of the gene, a process called *feedback inhibition,* or by environmental clues that the protein is not needed. Transcription may be enabled by the presence of a substrate for the enzyme to be produced. Enzymes that enable processes like transamination can be *allosterically* inhibited by the products of the reaction. Allosteric inhibition also occurs when the inhibitor binds to the DNA and changes its configuration such that the DNA can not be read out by the RNA. Also, high ADP levels may also inhibit enzymes required to produce proteins, diverting proteins to pathways that generate energy.

Regulation of protein synthesis is dependent on hormonal influences, including insulin, growth hormone, adrenaline, androgens, estrogens, progesterone, glucagons, glucocorticosteroids, and thyroid hormones. Protein synthesis is increased during growth periods and decreased during periods of fasting or physiological stress. During growth, muscle protein anabolism (synthesis) is especially sensitive to stimulation by dietary influence, apparently through the combined actions of insulin and the branched chain amino acids. Leucine, especially, at a very high dose can stimulate muscle protein synthesis, apparently due partially by stimulating the production of insulin. In conditions of physiological stress, such as injury and infection, synthesis declines and the organisms may be subject to muscle wasting. The reasons for this wasting are poorly understood. Research is ongoing to find ways to counter muscle wasting during prolonged illness or after surgery.

Summary

The nitrogen in proteins comes ultimately from atmospheric nitrogen. It is scavenged from the atmosphere by diazotrophic bacteria. These bacteria contain nitrogenase, which reduces the very stable, triple-bonded nitrogen to ammonia, a reaction that requires high energy input.

Once nitrogen is reduced, other living organisms can incorporate it into glutamate via an enzyme, glutamate synthetase. This enzyme replaces the carboxylic group on glutamate with an amide, creating glutamine. Glutamine can then donate the amide to various carbon skeletons ultimately producing the majority of the other amino acids. The movement of the amine requires aminotransferases and vitamin B_6 (pyridoxal phosphate). The majority of transaminations in humans occur to produce alamine or aspartate. Glutamate itself is produced from α-keto glutarate and glutamine by the enzyme glutamate dehydrogenase. This enzyme can act in reverse to release an amine and return the carbon skeleton to other metabolic processes. As such it has a key role in linking protein catabolism and anabolism.

Humans can produce 10 of the amino acids, which are termed nonessential amino acids and cannot produce the other 10, which must be consumed in the diet and are essential. Some of the nonessential amino acids, such as cysteine, require essential amino acids for their production, in this case methionine. If methionine availability is reduced, cysteine can become essential. Arginine is produced in the urea cycle and is normally needed only by infants. Aspartate can be consumed in the urea cycle is such quantities that it becomes essential.

The synthesis of many amino acids depends on folic acid (vitamin B_9) and also vitamin B_{12}, both of which enable transfer of single methyl groups between molecules. The structure of a component of folic acid, para aminobenzoic acid, is mimiced by sulfonamide drugs. These drugs are toxic to organisms that synthesize folic acid, such as bacteria. Humans take folic acid in their diet and therefore are not sensitive to sulfonamide drugs.

Amino acids are recycled in the body and also absorbed through the (gastro-intestinal) GI tract. About 400 g of protein is either recycled or degenerated every day. Only about 7% of this comes from the diet. The rest depends on metabolic process such as the glutamine-glutamate cycle, the alanine-pyruvate cycle, or the action of glutamate dehydrogenase moving nitrogen between glutamate and α-keto glutarate. Such processes move nitrogen between amino acids, move nitrogen from the peripheral tissues to the kidney and shunt the amino acid carbons into the TCA cycle.

Dietary proteins are digested in the GI tract by proteolytic enzymes, such as trypsin and chymotrypsin that break apart peptide bonds. Once inside the gastric cells, amino acids can be metabolized to enter energy-generating metabolic processes or to produce new proteins.

The urea cycle takes an amide from carbamoyl pyrophosphate and a second amide from another source, ammonia or glutamine, to convert aspartate into arginine, through a series of intermediates. The terminal carbon molecule, bearing two amine groups, is split off from arginine and forms the compound urea.

QUIZ

1. Without diazotrophs, the biosphere would lack
   (a) organisms with nitrogenase enzymes.
   (b) the ability to degrade organic material.
   (c) organisms that can live at high temperatures.
   (d) organisms that can intake plasmids.

2. Ferredoxin
   (a) is an oxygen transport molecule.
   (b) catalyzes nitrogen fixation.

    (c) stores electrons.

    (d) deaminates glutamate.

3. The production of essential amino acids in humans

    (a) occurs when amine is transferred to the corresponding α-keto acid.

    (b) cannot be synthesized by human metabolism.

    (c) starts with the creation of glutamate from α-keto glutarate.

    (d) starts with the creation of glutamine from glutamate.

    (e) None of the above.

4. If an essential amino acid is absent from the diet

    (a) the liver will produce the amino acid by transferring an amine from glutamine to the corresponding α-keto acid.

    (b) muscle protein will be degraded to scavenge the needed amino acid.

    (c) nitrogen will be conserved by recycling into glutamate or alanine.

    (d) glycogen will be degraded to scavenge the required α-keto acid.

    (e) Answers (b), (c), and (d) are correct.

5. Degradation of proteins into carbohydrates

    (a) requires proteases.

    (b) requires deamination.

    (c) is oxidative.

    (d) in some cases created intermediates that feed directly into the TCA cycle.

    (e) All of the above.

6. Transport of ammonium to the kidney can be done through

    (a) dissolved in serum.

    (b) conversion of glutamate to glutamine.

    (c) conversion of pyruvate to alanine.

    (d) All of the above.

7. In the urea cycle, the first amide group to enter the cycle

    (a) enters as ammonium ion.

    (b) is activated by attachment of a phosphate from the cleavage of ATP.

    (c) is provided by aspartate.

    (d) forms ornithine.

8. The following amino acids are derived from glutamate:

   (a) Serine.

   (b) Cysteine.

   (c) Methionine.

   (d) Arginine.

   (e) All of the above.

9. Biosynthesis of essential amino acids

   (a) requires exotic precursors not available to animals.

   (b) starts with common metabolic precursors.

   (c) can only be performed by diazotrophs.

   (d) can only be done by prokaryotic organisms.

   (e) Both (a) and (d) are correct.

10. Ubiquitin

   (a) attaches to proteins to mark them for destruction.

   (b) cleaves peptide bonds.

   (c) is found within proteosomes.

   (d) is secreted into the small intestine for protein digestion.

   (e) Both (b) and (d) are correct.

CHANGE 8

Nucleic Acids and Nucleotides

The *nucleotides* are one of the most important classes of molecules in the body. The first role nucleotides play is acting as the basis for sources of energy (adenosine triphosphate [ATP], guanosine triphosphate [GTP]) that drive biochemical reactions. They also are an important constituent in several coenzymes, but the most famous and dramatic role of the nucleotides is as a component of the nucleic acids, RNA and DNA.

In order to understand the nucleotides and nucleic acids and their role in the body, you will need to understand:

- Nitrogen bases, the pyrimidines and the purines
- How pyrimidine and purine rings are assembled
- The construction of RNA and DNA
- Basic function of RNA and DNA
- Degradation of the nucleic acids

Pyrimidines

The nucleotides are built from molecules called *nitrogen bases.* The first type of nitrogen base we consider are called *pyrimidines,* which include the following:

- Uracil
- Thymine
- Cytosine
- Orotic acid (relatively uncommon, but can be synthesized into uracil)

A pyrimidine is an organic molecule that has a six-membered ring with nitrogen atoms located at positions 1 and 3. For readers who have organic chemistry fresh in their minds, pyrimidine can be thought of as a nitrogen counterpart to a benzene ring. These are compared in Fig. 8-1.

Looking at Fig. 8-1, you can see that pyrimidine is an *aromatic* compound. This is because:

- It has a cyclic structure with alternating single and double bonds.
- There are three double bonds in the ring—giving six π-electrons.

You will recall that an *amine* is a molecule that has a nitrogen atom bonded to one, two, or three carbons. Since pyrimidine has three bonds from nitrogen to carbon, it is called a *tertiary amine.*

SYNTHESIS OF A PYRIMIDINE

A pyrimidine ring is synthesized from three components. These include:

- The amino acid *glutamine.*
- A compound called *carbomoyl phosphate.*
- And finally, *aspartate.*

The first step in the construction of a pyrimidine ring is the synthesis of carbomoyl phosphate from glutamine. This reaction is modulated by the enzyme *carbamoyl*

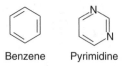

Benzene   Pyrimidine

Figure 8-1   A pyrimidine ring is similar to a benzene ring, except that it has nitrogen atoms located at positions 1 and 3.

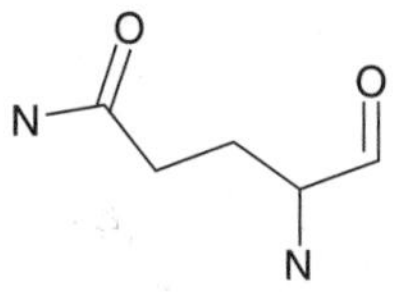

Figure 8-2   To make a pyrimidine ring, we start with the amino acid glutamine.

phosphate synthetase II (strangely abbreviated as *CAD* or sometimes *CPS II*) and takes place in the cytosol, primarily in the gastrointestinal (GI) tract, spleen, thymus, and testes. Glutamine, which is shown in Fig. 8-2, provides one of the nitrogen atoms of the pyrimidine ring.

The reaction, which requires 2 ATP molecules, proceeds as follows:

$$\text{Glu} + \text{HCO}_3^- + 2\text{ATP} \rightarrow \text{Carbamoyl phosphate} + \text{Glutamate} + 2\text{ADP} + \text{P}_i$$

Carbamoyl phosphate is illustrated in Fig. 8-3, where you can see that the ring structure is starting to take shape.

The next step is the formation of *orotic acid* (see Fig. 8-4), a compound which includes a six-membered ring structure with two nitrogen atoms. This reaction takes place in three steps:

- Step one: Carbamoyl phosphate + aspartate are transformed into the intermediary *N-carbamoylaspartate*. This reaction takes place in the presence of the enzyme *aspartate transcarbamylase*.

- Step two: *N*-carbamoylaspartate, in the presence of the enzyme *dihydro-orotase,* is converted into *dihydro-orotate* with the liberation of one water molecule.

- Step three: Finally, in the presence of the enzyme *dihydro-orotate dehydrogenase*, dihydro-orotate is transformed into orotic acid.

At this point we are almost done. Next, phosphoribosylpyrophosphate (PRPP) is added to orotic acid which results in the synthesis of *orotate monophosphate* (OMP). Uracil monophosphate (UMP) is then obtained from OMP by decarboxylation. Once UMP is in hand, production of the other pyrimidine bases is possible. It can be converted

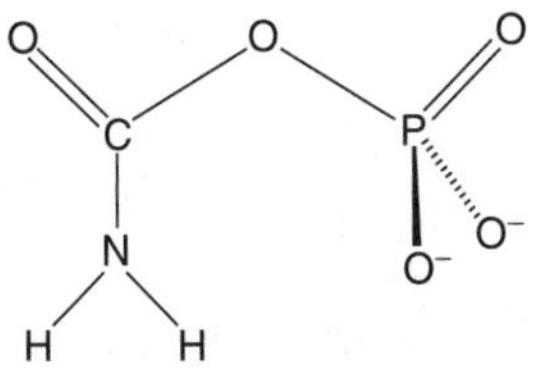

Figure 8-3   To make uracil, we start with glutamine which is converted to *carbamoyl-P*
in a reaction that requires 2 ATP molecules.

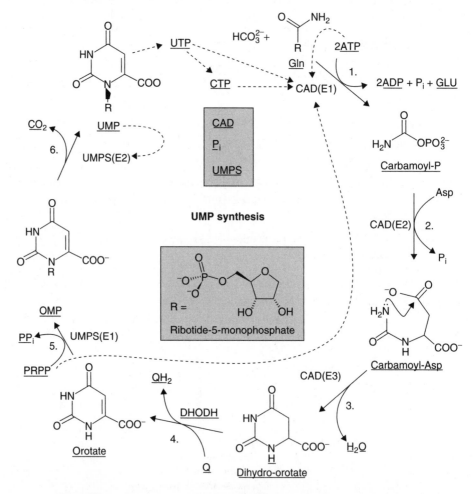

Figure 8-4 Orotic acid. An intermediary on the way to a pyrimidine ring.

into UTP which then yields cytosine tri-phosphate (CTP). The entire process is illustrated in Fig. 8-5.

Thymine is also derived from uracil. Methylation of the fifth carbon of uracil results in the production of thymine. The production process for the pyrimidine bases is summarized in Fig. 8-6.

Figure 8-5 The pathway for UMP synthesis. (*Courtesy of Wikipedia.*)

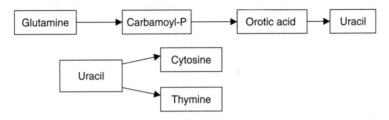

Figure 8-6   Uracil is synthesized from glutamine. It can then be used to make cytosine and thymine.

In Fig. 8-7, we show a pyrimidine ring and the origin of its constituents.

The three pyrimidine bases are shown in Fig. 8-8. Remind yourself that thymine and cytosine are derived from uracil.

Purine Synthesis

The nucleotides also include two *purine* bases. These are:

- Adenine
- Guanine

A purine is a bit more complicated than a pyrimidine. It consists of *two* nitrogen rings, one of which is a pyrimidine like we met in the last section. In addition, it contains an imidazole ring. This is a five-membered ring that contains two nitrogen atoms. An imidazole ring is illustrated in Fig. 8-9. A purine is illustrated in Fig. 8-10.

In the body, the biosynthesis of purines takes place primarily in the liver. The starting ingredient is PRPP which is converted to a purine called *inosine monophosphate* (IMP). Several intermediate steps are required which involve the addition of glycine, aspartate, glutamate, carbon dioxide, glucose, and ATP. This process is lengthy and complicated, so let's just dive in and summarize the steps required.

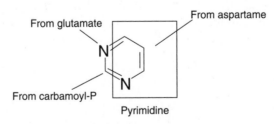

Figure 8-7   A pyrimidine ring consists of atoms from glutamate, carbamoyl-P, and aspartate.

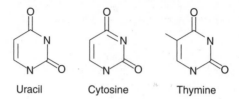

Figure 8-8   The pyrimidine bases uracil, cytosine, and thymine.

Step 1   The first step in the synthesis of a purine is the rate-limiting step. We begin with PRPP, replacing the pyrophosphate with the amide group of glutamine. The result is a compound called *5-phosphoribosylamine*. The reaction is catalyzed by the enzyme *Gln-PRPP amidotransferase*. Glutamate is a by-product of the reaction. This step is shown in Fig. 8-11.

Step 2   In the second step, the five-membered ring is constructed. We won't worry too much about the tedious details, except to note that 3 ATP, glycine, and glutamine are required to make the ring. The details along with some notes on required enzymes are shown in Fig. 8-12.

Step 3   The final step in the production of IMP is the synthesis of the six-membered ring. This step requires carbon dioxide, ATP, and aspartate. The details are illustrated in Fig. 8-13. Once IMP has been synthesized, it can be converted into AMP or GMP. IMP can be converted into either molecule. This is done through an intermediate molecule called *xanthosine monophosphate* or (*XMP*). IMP is oxidized by NAD giving XMP as illustrated here

$$NAD^+ \qquad\qquad NADH + H^+$$

$$IMP \longrightarrow XMP$$

Next, with the expense of one ATP and glutamine, we obtain GMP. This replaces a doubly bonded oxygen on XMP with an NH_2 donated from glutamine:

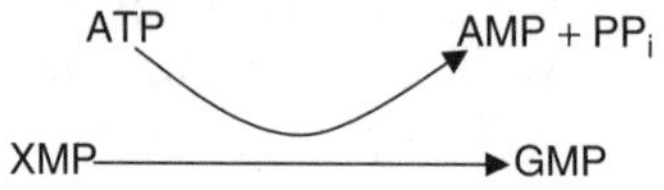

$$ATP \qquad\qquad AMP + PP_i$$

$$XMP \longrightarrow GMP$$

The resulting purine ring, guanine is illustrated below in Fig. 8-14.

The final purine ring of interest, *adenine* is formed as AMP from IMP in a two-step process, which requires one GTP and liberates one fumarate molecule. The purine adenine is shown in Fig. 8-15.

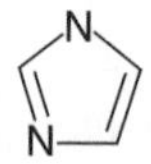

Figure 8-9   An imidazole ring is a component of a purine.

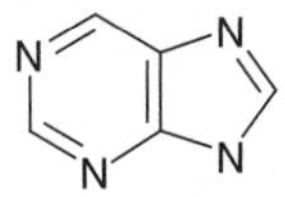

Figure 8-10   A purine consists of a pyrimidine and an imidazole ring. It contains four nitrogen atoms and four double bonds.

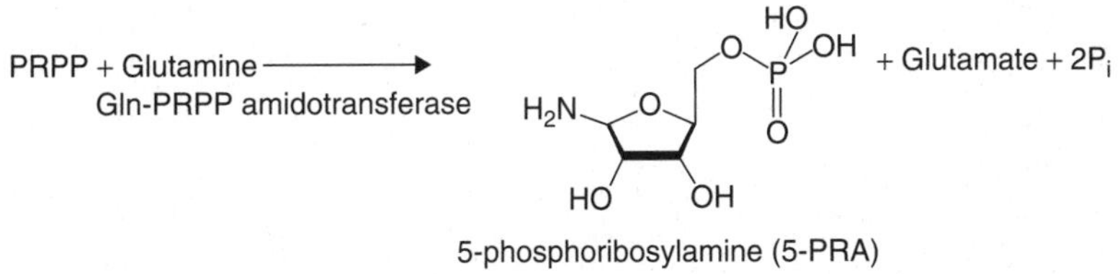

PRPP + Glutamine ──────────▶

Gln-PRPP amidotransferase

+ Glutamate + 2P$_i$

5-phosphoribosylamine (5-PRA)

Figure 8-11   The first step in the synthesis of a purine is the production of a molecule called 5-PRA.

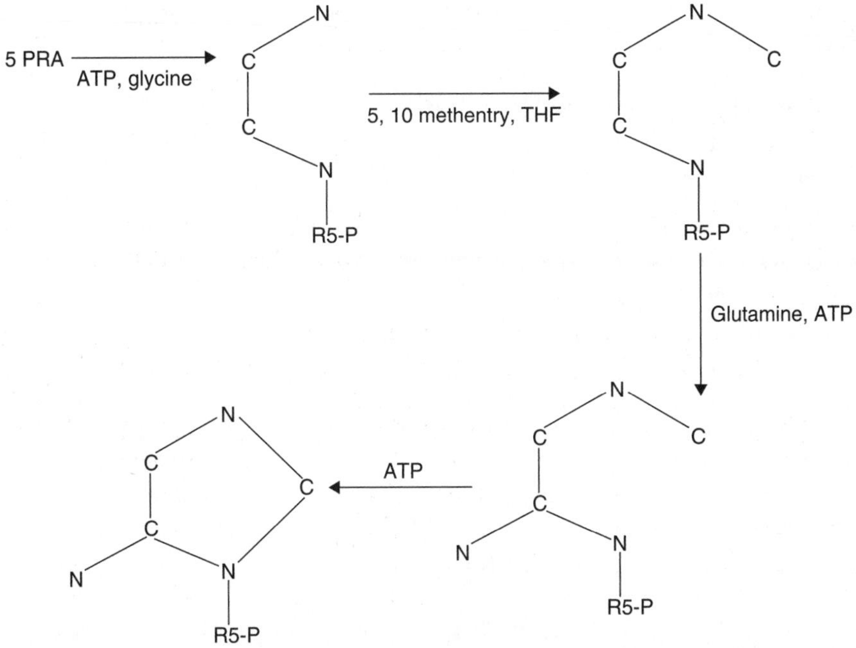

5 PRA ──────────▶
ATP, glycine

──────────▶
5, 10 methentry, THF

Glutamine, ATP

ATP ◀──────────

R5-P

Figure 8-12   The manufacture of the five-membered ring in the synthesis of purines.

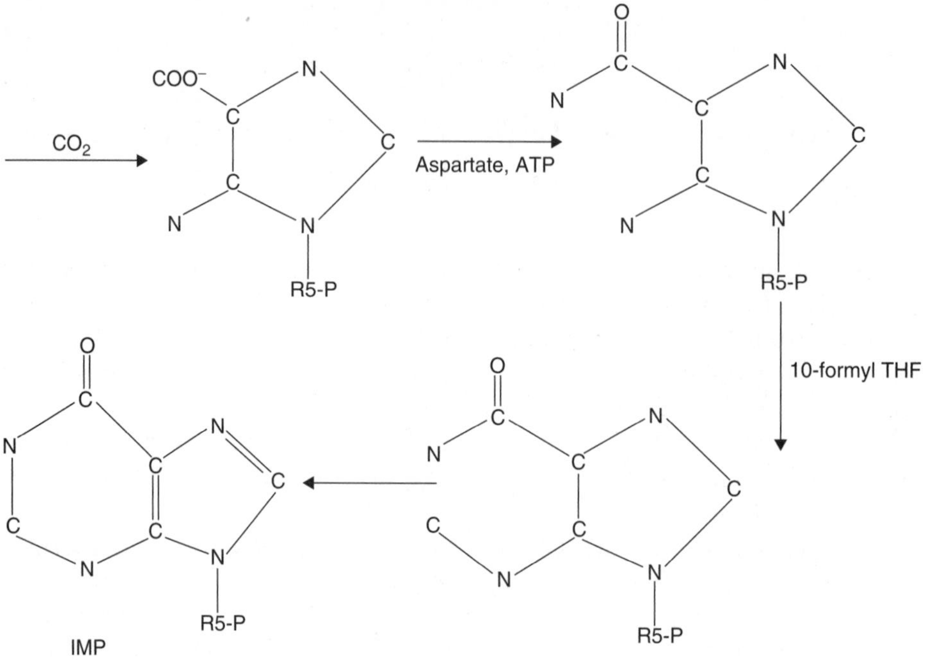

Figure 8-13   IMP is synthesized by the construction of a six-membered ring. The steps shown in this figure pick up where Fig. 8-12 leaves off.

Formation of the Nucleic Acids

Polymerization of the nucleotides produces the *nucleic acids*, the constituents of RNA and DNA. This reaction involves the liberation of a water molecule via the dehydration of an OH group at C-3′ of one nucleotide and at the phosphate group located at the C-5′ of another nucleotide. But first, how are the nucleotides built out of pyrimidine and purine rings?

Nucleotides are constructed from a phosphoric acid molecule, a sugar, and a purine or pyrimidine ring. The type of sugar in the molecule determines the type of nucleotide. If the sugar is ribose, we say that the nucleotide is a *ribonucleotide* and the resulting nucleic acid is *ribonucleic acid* (RNA). Ribose is illustrated in Fig. 8-16.

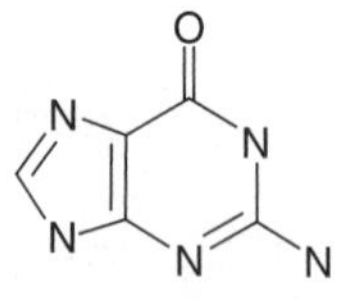

Figure 8-14   A guanine molecule, which is a purine ring.

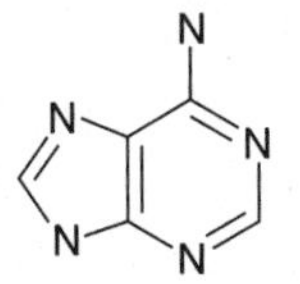

Figure 8-15   An adenine molecule.

When the sugar contained in the nucleic acid is *deoxyribose*, the resulting nucleic acid is DNA. Deoxyribose is shown in Fig. 8-17.

Deoxyribonucleotides are formed from the purine bases adenine and guanine and from the pyrimidine bases cytosine and thymine. This is done via a dehydration reaction that gives off 2 water molecules as shown in Fig. 8-18.

RNA molecules contain the sugar ribose and the bases adenine, guanine, cytosine, and uracil. Thymine cannot be contained in RNA because it will not form a nucleic acid compound with ribose.

The following figures show ball and stick models of each of the nucleic acids. In Figs. 8-19 to 8-22, we show the deoxyribonucleic acids thymine, cytidine, guanosine, and adenosine, respectively. Note the single-ring structures of thymine and cytidine (pyrimidines) and the double-ring structure of guanosine and adenosine (purines).

Within RNA base pairs are formed between guanine and cytosine, and adenine and uracil. In DNA, base pairs can be formed between guanine and cytosine, and adenine and thymine. Notice that uracil plays the role of thymine in RNA that thymine plays in DNA in base pair formation. This is an important distinction that influences the properties of the molecules. In Figs. 8-23 to 8-26, we show ball and stick models of uridine, cytidine, guanosine, and adenosine, the nucleic acid constituents of RNA.

The bases adenine and thymine play a role in the formation of the *double helix* structure of DNA. *Stacking forces,* which are forces along the length of a strand of nucleotides, differ in strength depending on which base is used. Thymine is found to have the weakest force between stacked units, while adenine has the strongest. Synthetic strands consisting solely of adenine will spontaneously form a coil. DNA has a double helical structure that is formed out of two strands that can be visualized as forming a twisted ladder or spiral staircase. The two strands are connected together by specific base pairings. These are A↔T and G↔C. The amount of purines (adenosine and guanosine) and the amount of pyrimidines is equal, due to

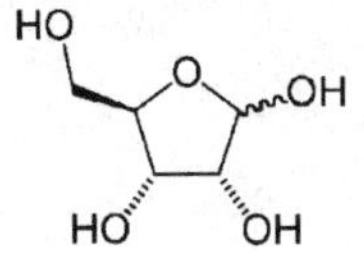

Figure 8-16   Ribose is the sugar component of RNA.

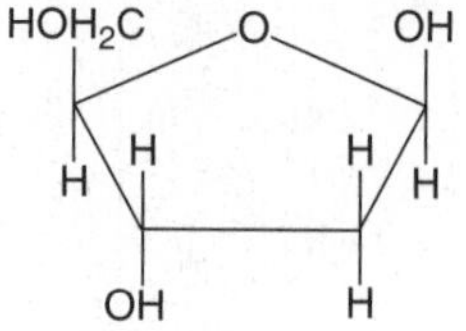

Figure 8-17 A deoxyribose molecule.

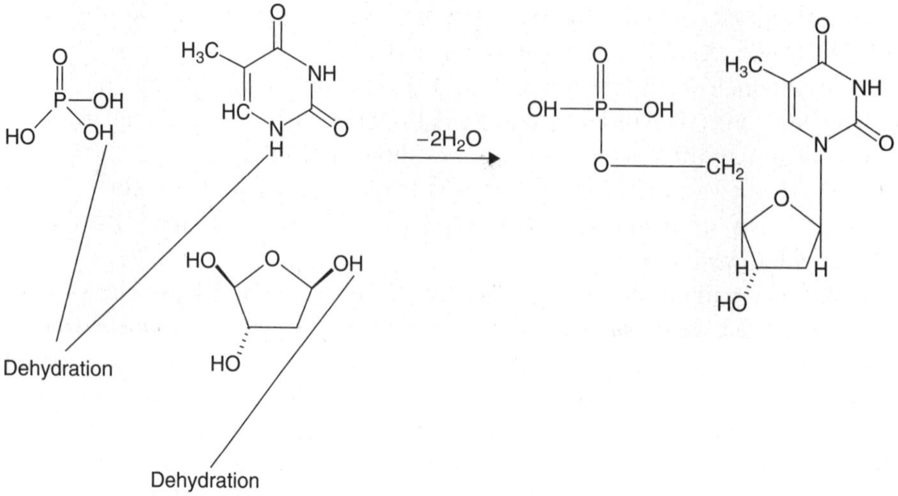

Figure 8-18 A dehydration process combines a base (in this case thymine) with a phosphate and deoxyribose sugar into a deoxynucleic acid molecule.

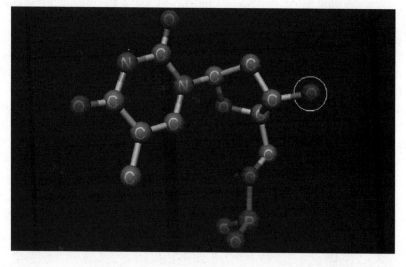

Figure 8-19 A ball and sick model of the DNA constituent thymine. Note that thymine is only found in DNA, and never in RNA.

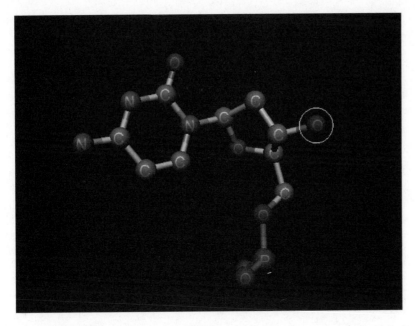

Figure 8-20   A ball and stick model of the nucleic acid cytidine.

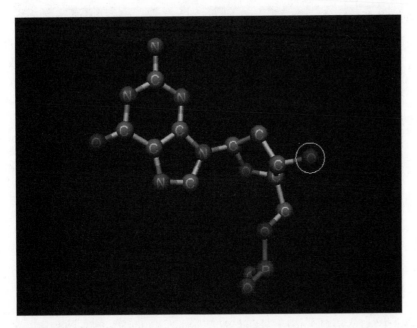

Figure 8-21   A ball and stick model of the nucleic acid guanosine.

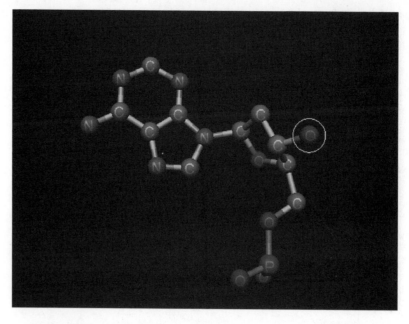

Figure 8-22   A ball and stick model of the nucleic acid adenosine.

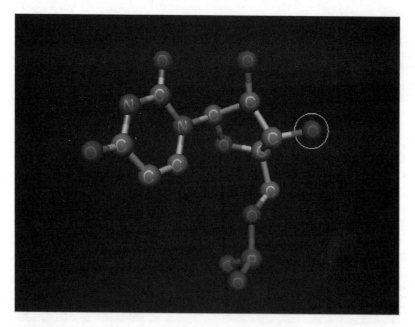

Figure 8-23   A ball and stick model of the nucleic acid uridine, found only in RNA.

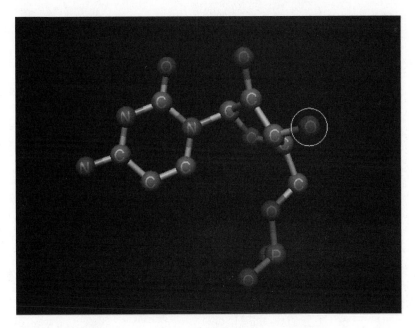

Figure 8-24   A ball and stick model of the nucleic acid cytidine, but with the sugar ribose instead of deoxyribose, as it is used in RNA.

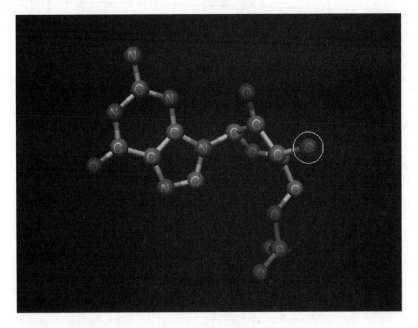

Figure 8-25   A ball and stick model of the nucleic acid guanosine, as found in RNA.

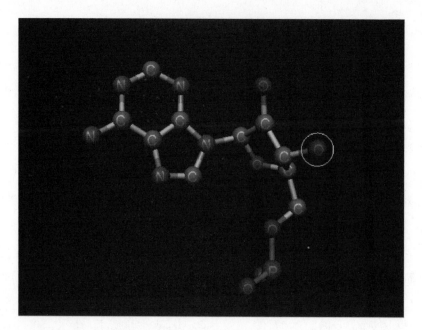

Figure 8-26   A ball and stick model of the nucleic acid adenosine as found in RNA.

the fact that only base pairings between a purine and a pyrimidine can occur. If a DNA strand has a higher G-C content, its stability is increased. The fact that the ratio of pyrimidines to purines is unity is also a reflection of the structural nature of the DNA molecule.

The double helix is formed through the base pairings by hydrogen bonds between the two complementary bases in a given pair. An A↔T pair has two hydrogen bonds, while a G↔C pair has three hydrogen bonds. The hydrogen bonding between adenine and thymine is shown in Fig. 8-27. The bonding between guanosine and cytosine is shown in Fig. 8-28.

Adenine   Thymine

Figure 8-27   Bonding between adenine and thymine in a DNA molecule involves two hydrogen bonds.

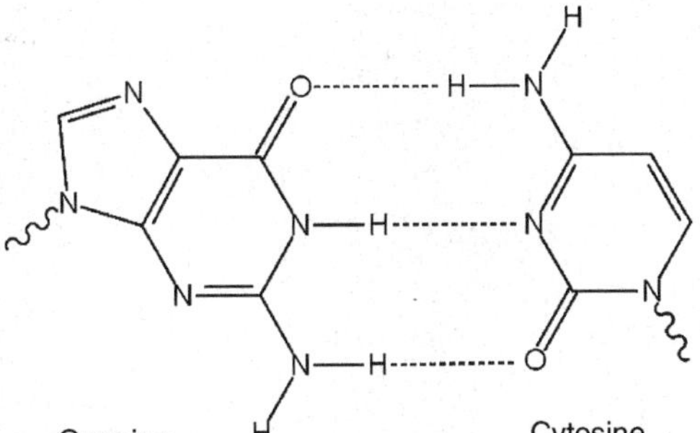

Figure 8-28   The bond between the base pairs guanine and cytosine in DNA involves three hydrogen bonds.

The two strands of DNA twist around each other giving what scientists call a *right-handed helix.* Each of the two individual chains is oriented in an opposite manner. Two ends of the chains defined by a phosphate and a hydroxyl group can be distinguished. We refer to the "top" end of the molecule, the end containing a phosphate as the 5′ (five prime) end, while the end with the hydroxyl group is called the 3′ (three prime) end. Looking straight on at the molecule, the strand on the left-hand side is shown with the 5′ end at the top and the 3′ end at the bottom. The strand on the right-hand side of the molecule must be in the opposite sense, so it has the 3′ end at the top and the 5′ end at the bottom. This is illustrated in Fig. 8-29. In Fig. 8-30, we show a three-dimensional representation of the DNA double helix.

DNA molecules are twisted or packed very tightly. Due to the twisting of the helix, the base pairs are brought very close to one another. There are hydrophobic interactions within the core of the strand which add to the strength given by the hydrogen bonds between pairs to make DNA a relatively stable molecule. Compared with the size of the cell, DNA is an extremely long molecule. A DNA molecule can contain up to 10^8 or 100 million base pairs. In a bacterial cell if its DNA were stretched out from end to end it would be 1000 times as long as the cell itself.

The composition of DNA, that is the relative ratios of each of the nucleotide bases, varies from species to species. For example, the G, A, C, T percent composition of the genome of *Escherichia Coli* is 24.9%, 26.0%, 25.2%, and 23.9% respectively. Now consider a human liver cell. Its G, A, C, T percent composition is found to be 19.9%, 30.3%, 19.9%, and 30.3% respectively. This is a reflection of the different genes that each organism has in its genome, which are composed of different nucleotides.

Now we turn to RNA, the simple minded cousin of the sophisticated DNA double helix. As noted above, the bases of RNA are adenine, guanine, uracil, and cytosine. Note the substitution of uracil for thymine-RNA can only contain bases that have

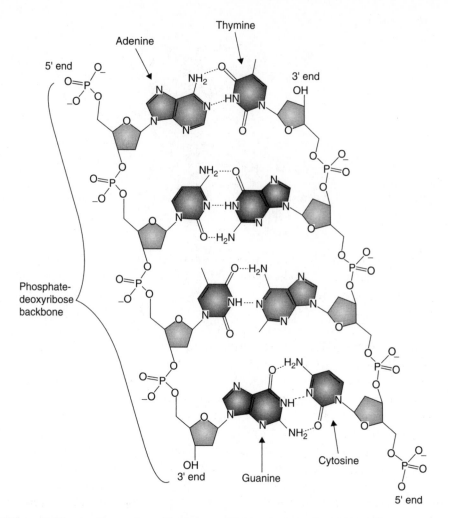

Figure 8-29   An illustration of laying two DNA strands end to end, with the concept of the 5′ end and the 3′ end. Note that the two strands run in opposite directions. (*Courtesy of Wikipedia.*)

ribose sugars. RNA can be found both in the nucleus and within the cytoplasm of the cell. While there is only one form of DNA, RNA comes in three basic varieties. Also, while DNA is found as the double helix in the nucleus, RNA is usually found in single stranded form. Note that although RNA exists as a single chain, regions of the chain can fold up on themselves forming double helical regions of an RNA molecule. This is done via A ↔ U and G ↔ C hydrogen bonding. About 50% of the nucleotides in an RNA molecule will form base pairs. Sections of RNA that do not form base pairs often form loops. tRNA is illustrated in Fig. 8-31.

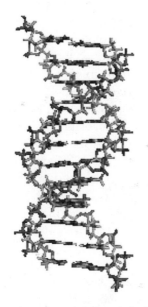

Figure 8-30   An illustration of a DNA double helix. (*Courtesy of Wikipedia.*)

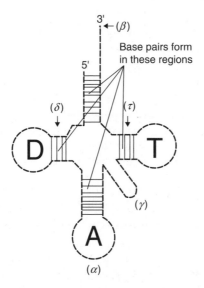

Figure 8-31   A tRNA molecule. Base pairs form in the regions indicated by crossed
lines. The loops in RNA are sometimes called *hairpin loops*.

Messenger RNA or mRNA is used to transcribe DNA in the nucleus so that the information needed to make a given protein can be transported from the nucleus out into the cytoplasm (the details of this process will be described in Chap. 14 on protein production). In the nucleus, the DNA double helix unwinds allowing RNA base pairs to form a single chain copy of a given nucleotide sequence. A segment of DNA is transcribed into a single strand segment called *pre-mRNA*. This strand includes sequences of nucleotides that do not code for any amino acids. We call these seemingly useless interruptions *introns*. The introns are removed from the pre-mRNA by enzymes, leaving behind sequences of nucleotides called *exons* which are spliced together to form a strand of mRNA that actually codes for a protein. The mRNA then moves out of the nucleus and into the cytoplasm to sites called the *ribosomes*.

It is here, in the cytoplasm, that we meet the second RNA sibling, *ribosomal RNA* or rRNA. Ribosomal RNA accounts for approximately 80% of the cells total RNA content. It resides in the ribosomes along with proteins.

The final variety of RNA we meet in this chapter is called *transfer RNA* or tRNA. About 10 to 20% of the cells total RNA content is tRNA. This type of RNA is involved with binding specific amino acids that are assembled together in the ribosomes to form proteins. Since there are 20 amino acids used in life processes, there are 20 types of tRNA molecules. While tRNA contains A, G, U, and C nucleotides, it also contains pseudouridine and inosine bases. tRNA is structured like a clover leaf, with base pairs forming pair bonds in many parts of the molecule. The "stem" end of the clover leaf is called the *acceptor* and it is here where we find the 3′ (OH group) end of tRNA on the left and the 5′ (phosphate) end on the right-hand side.

Remember, RNA contains a ribose sugar while DNA contains a deoxyribose. The lack of a hydroxyl group in the pentose ring of a DNA molecule helps give DNA a bit more stability than RNA, which is sensitive to hydrolysis. As a result, RNA contains some more complicated secondary structure to add stability in an attempt to compensate for the presence of the hydroxyl group. Moreover, the presence of the OH group at the C-2′ of ribose interferes with the formation of a long helix in RNA.

The differences between RNA and DNA can be summarized as

- DNA is found in a double helix structure, while RNA is usually found as a single strand.
- DNA contains thymine, but RNA contains uracil.
- DNA contains a deoxyribose sugar, while RNA contains a ribose.

Denaturation, Damage, and Degradation of DNA and RNA

DNA can be *denatured* by heating a sample in solution to 95°C. This causes the double helix structure to collapse. The two-stranded molecule separates into two simple single strands. Denaturation of DNA can also occur via any reagent that will weaken hydrogen bonding between base pairs or by any reagent that decreases the polarity of the solution (remember that the double helix gets some of its strength from hydrophobic regions). DNA that is denatured will spontaneously reassemble when temperatures are lowered in a process called *annealing*.

A *mutagen* is an agent capable of damaging DNA and can result in the alteration of a given gene. Mutagens are found in the environment and include chemicals such as benzene and free radicals, radiation such as UV light and x-rays, and some viruses.

In a process called *intercalation*, a molecule inserts itself in between base pairs in the DNA double helix. Aromatic rings such as benzene can damage DNA in this way. A free radical, on the other hand, can cause breaks in the DNA strand or even alter the specific base pairs in a DNA sequence. If such damage is not repaired, this can cause damage to a given gene.

UV light damages the DNA in a process that forms *thymine dimers*. This results in the formation of bonds between pyrimidine bases (between adjacent thymines). Remember that in a normal DNA double helix, bonds only form between pyrimidine-purine pairs, so any bonds between pyrimidine bases are defective. Thymine dimers can cause problems in DNA replication and transcription.

Damaged DNA can be repaired in the cell. This is done using a process called an *excision repair,* in which a damaged region of DNA is excised and a new segment is put in its place. This process involves several enzymes, and will be discussed in Chap. 9 on enzymes.

A great deal of postprocessing gene regulation goes on in the cell. This can be done by RNA *degradation*. The degradation of mRNA can regulate the amounts of various proteins found in the cell. One mechanism by which this is accomplished is by the binding of two proteins, *Staufen 1* and *Upf1* to an RNA molecule. The RNA molecule is subsequently degraded. Cell signaling via phosphorylations of the two proteins regulates their activities in the cell. In the nucleus sometimes aberrant mRNA transcripts appear. By a process called *polyadenylation,* the 3′ end of these mRNA strands gets marked (like the scarlet letter) and the mRNA strands are singled out for destruction.

Division of DNA

The strategy for copying the information stored in one cell's chromosomes and passing the information to two new cells in its entirety is very simple: unwind the DNA and break apart the hydrogen bonds that hold the base pairs together. At this point, The DNA exists as two single strands with organic bases protruding from the sugar backbone. A given base always pairs with the same partner, adenine with thymine and cytosine with quanine a phenomena known as *complementary base pairing*. Therefore, each base on each strand will pick up its partner in the form of a nucleotide out of the nucleoplasm in a reaction catalyzed by the enzyme *DNA polymerase*.

The sugars and the phosphates of the new, juxtapositioned nucleosides will join together by a phosphodiester linkage to form the backbone. Voilà! You now have two identical chromosomes where once there was one . . . except, it's a little more involved than just that.

For the cell to begin the process of duplicating the chromosomes, unzipping the chromosome is a bit of a challenge. Most of us think of chromosomes as long, winding staircases, because that is how we have always seen it depicted. Actually, even in a resting cell, the DNA material is wound very, very tightly, staircase on top of staircase. Consider the fact that the DNA inside a single cell, if laid end to end, would be over an inch long. The DNA is intertwined like many strands of hemp forming a rope. In eukaryotic cells, the DNA rope winds around proteins balls called *histones*. The DNA-histone complex twists to form flat discs.

Before the DNA can be replicated, it must be "unwound" by the enzyme *helicase*. Then the hydrogen bonds are broken, the stands are pulled apart, new strands are formed, and the material condenses into two separate chromosomes.

Cell division is tightly controlled. Certain cells divide frequently, such as bone marrow cells; certain cells divide primarily in growing or injured organs, such as bone cells; and certain cells divide rarely, if at all, such as mature nerve cells. Uncontrolled growth is the hallmark of cancer. The division process requires something to kick it off, and this "something" is called an *activation factor*. An example of an activation factor is the human growth factor.

The way activation factors usually work is by altering the shape of the DNA, allowing the enzyme that performs the division of the chromosomes, DNA polymerase, to bind. The DNA polymerase first finds a specific shape on the DNA. This shape, a loop, is required before the polymerase can bind to the DNA and do its thing. This looped configuration is assumed when the activation factor binds to the DNA. The area where the DNA polymerase initiates DNA replication is called the *origin of replication*. The DNA polymerase requires a *primer*, composed of specialized RNA molecules before the duplication can begin. Once the DNA polymerase finds a binding site, it chugs up the chain, separating the strands and building new ones by adding complementary base pairs to the single strands of DNA.

One of the most important jobs of the DNA polymerase is to correct errors in the newly synthesized DNA. When an incorrect base pair is recognized, the DNA polymerase reverses its direction by one base pair of DNA. The incorrect base pair is excised and the correct one inserted in an activity known as *proofreading*. As a result, an error in DNA replication is propagated only about 1 in each 10 to 100 billion base pairs. However, because cells divide so frequently, mistakes are inevitably transmitted. Many of the mistakes result in cells that are not viable. If a mutated cell does survive, the immune system has cells that ferret out and destroy these abnormal cells.

Summary

The nucleotides are formed from derivatives of two fused ring compounds called pyrimidines and purines. A pyrimidine is a single six-membered ring while a purine consists of two rings, a five-membered and a six-membered ring. The pyrimidines include the bases uracil, thymine, and cytosine, and are synthesized from glutamine, carbamoyl phosphate, and aspartate. The purines include adenine and guanine and their synthesis requires an imidazole ring. Nucleic acids are synthesized from these pyrimidine and purine bases with the addition of a phosphate group and a sugar. In the case of RNA, the sugar used is a ribose, in the case of DNA, the sugar is a deoxyribose. The sugar in RNA has an OH group which has a large influence on the structure of RNA by preventing it from forming long helixes. As a result RNA usually exists as a single stranded molecule. It comes in three varieties, messenger RNA, ribosomal RNA, and transfer RNA, and RNA is used to synthesize proteins.

DNA, the blueprint of life, resides in the nucleus of the cell. It includes the bases G, A, C, T which form base pairs via hydrogen bonds as A ↔ T and G ↔ C. In RNA, uracil is substituted for thymine because a nucleic acid containing thymine and a ribose sugar is not possible. RNA is found both in the nucleus and in the cytoplasm.

DNA divides by breaking hydrogen bonds, separating the double strands and adding complementary bases to the organic bases on the two single strands. The separation of the strands and the addition of the complementary bases is accomplished by the enzyme DNA polymerase. DNA polymerase also corrects inevitable mistakes made in reassembling the double strand. The proofreading by DNA results in an almost mistake-free process with misreads every 10 billion bases added.

Quiz

1. Which of the following is an aromatic ring?

   (a)

   (b)

   (c)

   (d)

2. Which of the following is not a pyrimidine?

   (a) Thymine

   (b) Orotic acid

   (c) Guanine

   (d) Cytosine

   (e) Uracil

3. The synthesis of a pyrimidine requires

   (a) glutamine, carbamoyl sulfate, and aspartate.

   (b) glutamine, carbamoyl phosphate, and aspartate.

   (c) glycine, carbamoyl phosphate, and aspartate.

   (d) an imidazole ring, glutamine, carbamoyl phosphate, and aspartate.

4. A pyrimidine that can be used to make other pyrimidine bases is

   (a) uracil.

   (b) thymine.

   (c) guanine.

   (d) adenosine.

5. Orotic acid is synthesized into

   (a) cytosine.

   (b) imidazole.

   (c) thymine.

   (d) uracil.

6. Which of the following is not a true statement regarding the difference between DNA and RNA?

   (a) RNA contains the nucleic acids adenosine, guanosine, cytidine, and uridine, while DNA only contains thymine.

   (b) In DNA, the base thymine is substituted for uracil which is found in RNA.

   (c) The sugar in RNA is ribose which has a hydroxyl group that interferes with the formation of long helical structures, while DNA contains the sugar deoxyribose.

   (d) DNA is found in a helical structure, while RNA is usually single stranded.

7. In the DNA double helix

   (a) guanine and cytosine bond with 2 hydrogen bonds, while adenine and thymine bond with 3 hydrogen bonds.

   (b) guanine and cytosine bond with 3 hydrogen bonds, while adenine and thymine bond with 2 hydrogen bonds.

   (c) guanine and cytosine bond with 3 hydrogen bonds, while adenine and thymine bond with 3 hydrogen bonds.

   (d) guanine and cytosine bond with 2 hydrogen bonds, while adenine and uracil bond with 3 hydrogen bonds.

8. The formation of a nucleic acid involves

   (a) a dehydration process involving a base, a phosphate, and a pentose sugar.

   (b) a hydrolysis process involving a base, a phosphate, and a pentose sugar.

   (c) a dehydration process involving a base, a phosphate, and a hexose sugar.

   (d) None of the above.

9. The process of annealing is

   (a) the spontaneous reassembly of DNA after degradation.

   (b) the fusing of two strands of RNA.

   (c) the spontaneous reassembly of DNA after denaturation.

   (d) None of the above.

10. UV light damages DNA by

   (a) causing two purine rings to bind together.

   (b) causing the formation of thymine dimers.

   (c) allowing an aromatic ring to insert itself in between two base pairs.

   (d) causing the double helix to split apart.

CHAPTER 9

Enzymes

An *enzyme* is a molecule that *catalyzes* or greatly speeds up the rate at which a chemical reaction will occur. In most cases, enzymes are proteins, but some enzymes, called *ribozymes,* are made of RNA. The key aspect of an enzyme is that it reduces the activation energy required for a reaction. In doing so, it speeds up the reaction. The factor of speedup can be very large indeed, up to 10^{20} in some cases. From this we can see that enzymes are vital in biological processes. Without enzymes, many chemical reactions in the cell which make life possible would probably not even occur.

We call molecules that participate in a reaction moderated by an enzyme *reactants* or *substrates*. We will use both terms interchangeably. The end products of the reaction are called *products*. During a reaction, the enzyme binds to one or more reactants, and it is the binding of the enzyme to the reactant that lowers the activation energy. It is truly amazing and fortuitous that this process occurs in nature allowing life to exist.

We call the site to which a substrate binds to the enzyme the *active site*. A substrate can become bound to an enzyme via one of the following ways:

- Hydrogen bonding
- Van der Waals interactions
- Electrostatic attraction

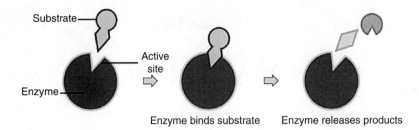

Figure 9-1 In the lock and key model, the substrate fits the active site of the enzyme the way a key fits a lock. (*Courtesy of Wikipedia.*)

Binding of a substrate to an enzyme can be surprisingly specific. In fact, in some cases only one stereoisomer will bind to the enzyme (explaining why some stereoisomers are biologically inactive). To understand the binding process, two analogies are used to describe it. The first is called the *lock and key* model. In this case, we view the substrate as a *key* and the enzyme as a *lock* to which it must fit. A given key can only open a specific lock and each key/lock combination has a mechanical fit. In this way, we can visualize the substrate as fitting the active site of an enzyme just like a key would fit a lock. This is illustrated in Fig. 9-1.

In some cases, the substrate and the enzyme do not have an immediate fit. Rather, the substrate binds to the enzyme by *inducing* the enzyme to assume a structural shape such that there is a lock and key fit. We call this type of binding the *induced fit model*. The induced fit model is illustrated in Fig. 9-2.

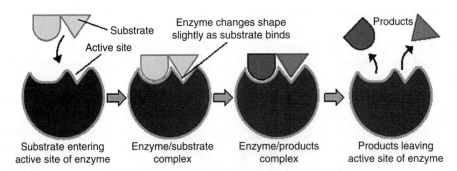

Figure 9-2 Sometimes an enzyme and substrate do not have an exact fit, but the presence of the substrate induces the enzyme to change shape so that the fit is exact. We call this model of enzyme substrate interaction the induced fit model. (*Courtesy of Wikipedia.*)

Enzyme Naming Conventions

In most cases, a naming convention is used for enzymes such that the suffix ASE is appended to the name of the substrate associated with the enzyme. For example,

- LACTASE: Breaks down the sugar lactose.
- HEXOKINASE: Converts glucose (a hexose) into glucose 6-phosphate.
- ALCOHOL DEHYDROGENASE: Converts ethanol into acetaldehyde.

Not all enzymes are named using this convention, some compounds have older names such as *trypsin*, an enzyme found in the digestive system that breaks down proteins.

FACTORS INFLUENCING ENZYME ACTIVITY

When you think of enzymes think of *speeding up a reaction.* How well an enzyme works determines how much it can speed up a reaction. There are several factors that can influence how well an enzyme works. These are

- Temperature
- pH of the local environment
- Concentration of the substrate
- Presence of *inhibitors*
- Presence of *cofactors* and *coenzymes*

Each of these factors will be discussed throughout the chapter. Now we turn to a description of enzymes and reaction rates.

Enzyme Kinetics

A description of how an enzyme impacts the reaction rate is called *enzyme kinetics.* Whether a given reaction can occur or the direction in which the reaction goes is determined by the change in standard free energy $\Delta G°$ for the reaction. The *rate* of the reaction—how fast it goes—is determined by the activation energy $\Delta G^{°\ddagger}$. Enzymes work by lowering $\Delta G^{°\ddagger}$.

If the concentration of substrate molecules is high, an enzyme will work faster. This is because the enzyme will be colliding with substrate molecules more often and will bind with them more often as a result. In addition, we know that temperature is a measure of molecular speed. At higher temperatures, molecules will be moving around in solution more rapidly, increasing the probability for a collision. Hence, at higher temperatures you can expect that an enzyme-mediated reaction will progress faster. In addition, at higher temperatures individual molecules have more energy;

hence, more molecules will have enough energy to match the activation energy for a given reaction.

However, the increase in reactivity with temperature only works to a point. All enzymes have an upper limit in temperature above which the reaction will not proceed at a faster rate. This is because at high temperatures proteins will denature, that is, the structure of the enzyme will break down and it will no longer work like it does under normal conditions. But under normal conditions the body is under constant temperature and pressure, so we will not concern ourselves with this issue in detail.

Enzyme activity is also affected by *pH* and the presence of *inhibitors,* molecules which inhibit an enzyme-mediated reaction. Enzymes increase the rate of a reaction over a narrow region of pH, and outside this narrow range the activity of the enzyme rapidly drops off. We will discuss inhibitors later in this chapter.

Changes in Free Energy and Activation Energy

The change in free energy ΔG° is the difference in the free energy of the reactants and the free energy of the products. For a reaction to be favored, the free energy of the products should be lower than the free energy of the reactants. Let G_R be the free energy of the reactants and G_P be the free energy of the products. Then, for a reaction to be favored we should have

$$\Delta G^\circ = G_P - G_R < 0$$

The activation energy $\Delta G^{\circ\ddagger}$ is the difference between the peak free energy during the reaction and the free energy of the reactants (see Fig. 9-3). The affect of an

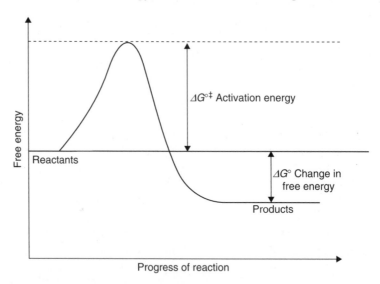

Figure 9-3   An illustration of the activation energy and change in free energy in a reaction.

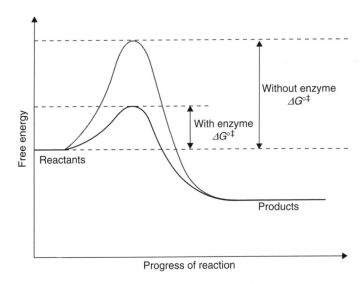

Figure 9-4   An enzyme lowers the activation energy for a reaction.
(Courtesy of Wikipedia.)

enzyme is to lower the maximum free energy G attained during the reaction, which in turn lowers the activation energy $\Delta G^{\circ\ddagger}$. This is illustrated in Fig. 9-4, where we compare the activation energy for a reaction taking place *without* the presence of an enzyme and in the presence of an enzyme.

The way that an enzyme mediates a reaction can be illustrated in the following way. For simplicity, we consider only a single substrate S and single product P.

Let E be the enzyme. The first step in an enzyme-mediated reaction is the binding of the enzyme to the substrate giving the *enzyme substrate complexes*. The binding process is reversible so that we have

$$E + S \rightleftharpoons ES$$

The second step is called the *catalytic step*. In this step, the enzyme catalyzes the given reaction $(S \rightarrow P)$ and releases the product. So, in the presence of the enzyme

$$ES \rightarrow E + P$$

Characterizing Enzyme Kinetics

The *reaction velocity* is the number of reactions that are catalyzed by an enzyme per second. As the concentration of a substrate is increased, the reaction rate increases until it reaches a *saturation point*. This is the maximum rate at which a reaction can

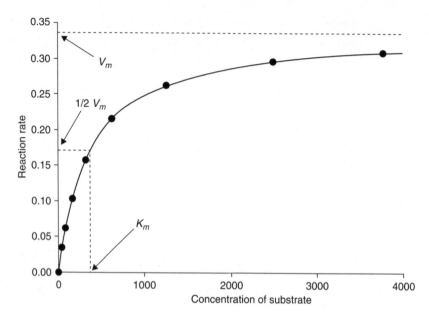

Figure 9-5 A saturation curve. At the saturation point, the reaction is proceeding at the maximum velocity V_m. (*Courtesy of Wikipedia.*)

occur, because all enzyme molecules are bound to substrates. When the reaction is completed, an enzyme molecule releases the product and immediately binds another substrate molecule. We denote the maximum rate at which the reactions will occur by V_m. A plot of the concentration of a substrate versus the reaction rate is called a *saturation curve,* which is illustrated in Fig. 9-5.

An enzymes affinity for a given substrate can be characterized by the point at which the reaction velocity is one-half the maximum velocity V_m for the reaction. We denote the concentration of substrate for which this occurs by K_m and at this concentration of substrate $V = \frac{1}{2}V_m$. We call K_m the *Michaelis-Menten constant.*

Now, thinking in terms of a general chemical reaction (not necessarily catalyzed by an enzyme), consider a reaction of the form:

$$A + B \rightarrow P$$

Then the *rate* of the reaction can be characterized by the change in concentration of the product with respect to time. That is:

$$\text{Rate} = \frac{\Delta P}{\Delta t} \rightarrow \frac{dP}{dt}$$

In the limit of infinitesimal time changes, this becomes a derivative. The rate of the reaction is also proportional to the decrease in the concentrations of the reactants with time. That is:

$$\text{Rate} = -\frac{\Delta A}{\Delta t} = -\frac{\Delta B}{\Delta t}$$

In general, the rate of the reaction is proportional to the concentrations of the reactants in the following way:

$$\text{Rate} = k[A]^x [B]^y$$

The constant of proportionality k is called the *rate constant* for the reaction. If x and y are integers, then the *order* of the reaction is $x + y$. The units of k depend on the order of the reaction.

Now, let k_1 be the rate constant for the reaction in which the enzyme and substrate bind together:

$$E + S \xrightarrow{k_1} ES$$

Let k_{-1} be the rate constant for the reaction in which the enzyme and substrate dissociate without the catalyzed reaction taking place:

$$ES \xrightarrow{k_{-1}} E + S$$

Finally, let k_2 be the rate constant for the enzyme-substrate to form the product

$$ES \xrightarrow{k_2} E + P$$

We can characterize the velocity of an enzyme catalyzed reaction (i.e., the rate) at any point in the following way:

$$V = \text{reaction rate} = k_2[ES] = \frac{V_m[S]}{\dfrac{(k_{-1} + k_2)}{k_1} + [S]}$$

This is called the *Michaelis-Menten equation*. Now define

$$K_m = \frac{(k_{-1} + k_2)}{k_1}$$

Then:

$$V = \text{reaction rate} = k_2 [ES] = \frac{V_m [S]}{K_m + [S]}$$

Let's examine the behavior of this expression in a couple of important cases. Notice that if the concentration of the substrate is much larger than K_m, meaning that it is much larger than the point at which the reaction rate is one-half of maximum $V = \frac{1}{2} V_m$, we have $[S] \gg K_m \Rightarrow K_m \approx 0$ and so

$$V = \frac{V_m [S]}{K_m + [S]} \approx \frac{V_m [S]}{[S]} = V_m$$

Now, if the concentration of the substrate is at K_m, then we have

$$V = \frac{V_m [S]}{K_m + [S]} = \frac{V_m K_m}{K_m + K_m} = V_m \frac{K_m}{2 K_m} = \frac{1}{2} V_m$$

Finally, if the concentration of the substrate is very small, that is $K_m \gg [S]$, then we can take $K_m + [S] \approx K_m$, therefore

$$V = \frac{V_m [S]}{K_m + [S]} = V_m \frac{[S]}{K_m}$$

Affinity of an Enzyme for a given Substrate

The Michaelis-Menten constant K_m describes the *affinity* of an enzyme for a given substrate. If K_m is small, then the enzyme has a high affinity for the substrate. A small K_m means that the reaction rate will reach one-half the maximum at very small concentrations of the substrate. If K_m is small, the enzyme substrate complex ES is tightly bound and the reaction $ES \rightarrow P$ will take place before the enzyme-substrate complex has a chance to dissociate ($ES \rightarrow E + S$).

On the other hand, if K_m is large, the ES complex is not tightly bound. This means that the enzyme-substrate will tend to dissociate ($ES \rightarrow E + S$) before the reaction $ES \rightarrow P$ has a chance to proceed. Another way to look at this is that if K_m is large, high concentrations of the substrate are required for the enzyme to catalyze the reaction. Summarizing these important points

- If K_m is small, the enzyme and substrate are tightly bound. The reaction is likely to proceed without them dissociating. Small concentrations of substrate are necessary for the reaction to reach a rate of one-half the maximum.

- If K_m is large, the enzyme and substrate are loosely bound. They may dissociate before the reaction takes place. Large concentrations of substrate are necessary for the reaction to reach a rate of one-half the maximum.

The Michaelis-Menten constant K_m is important in many bodily processes. This is true in the case of glycolysis, where different enzymes have different affinities for the glucose molecule. Specifically, the K_m for binding to glucose in the case of the enzymes hexokinase and glucokinase is

$$K_m(\text{hexokinase}) = 0.15 \text{ mM}$$
$$K_m(\text{glucokinase}) = 20 \text{ mM}$$

Since $K_m(\text{glucokinase}) \gg K_m(\text{hexokinase})$, we know that glucose binds to glucokinase in a rather loose fashion. Hexokinase has a high affinity for glucose; glucokinase has a low affinity for glucose. So it is unlikely that glucose will stay bound to glucokinase long enough for the first reaction in glycolysis to proceed.

The Michaelis-Menten constant K_m can also be used to characterize the affinity of an enzyme for different substrates. Considering only the enzyme hexokinase, we can compare its affinity to glucose and fructose. In this case

$$K_m(\text{hexokinase-glucose}) = 0.15 \text{ mM}$$
$$K_m(\text{hexokinase-fructose}) = 1.5 \text{ mM}$$

This tells us that hexokinase has 10 times the affinity for glucose as it does for fructose. However, notice that hexokinase has a great deal more affinity for fructose than glucokinase does for glucose.

Turnover Number

The final quantity we will consider in our examination of enzyme kinetics is called *turnover number*. This is the number of reactions catalyzed by the enzyme per unit time, and we denote it by k_{cat}. The *efficiency* of an enzyme can be estimated by dividing the turnover number by the Michaelis-Menten constant.

$$\text{efficiency} \propto \frac{k_{cat}}{K_m}$$

The turnover number gives us the rate constant when the enzyme is saturated. An enzyme that is more efficient will have a larger turnover number, meaning that it can catalyze more reactions in a given time span.

TYPES OF ENZYMES

Enzymes can be classified by their mechanism of action or by what they do to the substrates. There are six major classes of enzymes. These include

- Transferase: This type of enzyme transfers a functional group from one substrate to another. Let's denote the functional group as X and denote the substrates as A and B. Then the reaction catalyzed by a transferase can be characterized as $A - X + B \rightarrow A + B - X$.

- Ligase: This type of enzyme joins two molecules together by generating a chemical bond between them. A ligase requires an energy molecule like ATP and the reaction is accompanied by hydrolysis.

- Isomerase: This type of enzyme, as its name implies, catalyzes a spatial rearrangement of the substrate molecule.

- Lyase: This is an enzyme that catalyzes the nonhydrolytic cleavage of single chemical bonds, leaving behind double bonds or a ring structure.

- Hydrolase: Catalyzes the hydrolysis of a chemical bond.

- Oxidoreductase: catalyzes the transfer of one or more electrons from a hydrogen acceptor or electron donor to a hydrogen donor. A reaction of this type is $A^- + B \rightarrow A + B^-$.

TYPES OF CATALYSIS

There are different means by which an enzyme can catalyze a reaction. We now review each of these in turn.

Covalent Catalysis

In this case, the enzyme and the substrate form a temporary covalent bond. This is possible because many substrates have electron deficient parts we call *electrophilic* (electron loving). In turn, many enzymes are large protein molecules containing amino acids with *nucleophilic* groups. That is, the enzyme contains amino acids that are negatively charged because they have unshared electrons or are simply electron rich. The bond formed between the enzyme and the substrate results in a *reaction intermediate* or *transition state*. The intermediate is then converted into the end product. This type of catalysis is often seen in transferase enzymes, where the flow of electrons during the reaction is indicated.

Acid-Base Catalysis

When a catalyzed reaction involves the transfer of protons, we say that the type of enzyme-mediated reaction is *acid-base catalysis*. This type of catalysis may take place in conjunction with other types of mechanisms taking place in the active site of the enzyme. This is necessary because acid-base catalysis by itself does not speed up the reaction rate all that much, maybe by a factor of 100. Acid-base catalysis can be either *general* or *specific*. The types of catalysis in this category include:

- General acid catalysis
- General base catalysis
- Specific or concerted acid-base catalysis

For *general acid catalysis*:

- The active site of the enzyme must have residues or side chains that are *protonated*. The effectiveness of the side chain as a catalyst will depend on the pH and pK_a of the environment.
- Protons are donated to the substrate.
- Amino acid side chains of glutamic acid, histidine, aspartic acid, lysine, tyrosine, and cysteine can act as acid catalysts when their amino acid side chains are protonated.

In concerted acid-base catalysis, both an acid and a base participate in the reaction.

Metal Ion Catalysis

When an enzyme requires the presence of a metal ion to do its job we call this *metal ion catalysis*. If the metallic ion is loosely bound to the enzyme, we say that the enzyme is *metal activated*. Some metallic ions typically seen in metal activated enzymes are Na^+, K^+, and Ca^{2+}.

A *metalloenzyme* is a compound consisting of an enzyme which is tightly bound to a metallic ion which acts as a cofactor (see below). Some metallic ions involved in the activity of metalloenzymes include Fe^{2+}, Cu^{2+}, and Zn^{2+}. Metalloenzymes are heavily influenced by pH level due to its effect on electron flow.

Approximately one-third of all enzymes use metal ions. The reason is that the metal enzymes can be used in electron transfer. As an example, consider *carboxypeptidase A*, an enzyme involved in the digestion of proteins. It is a metalloenzyme containing zinc.

SPECIFICITY

How *specific* an enzyme is to a given substrate can be classified in one of three ways. These are

- *Absolute:* In this case, an enzyme only works on one substrate producing one product.

- *Relative:* If an enzyme has relative specificity, it will work with several structurally similar substrates, catalyzing reactions producing structurally similar products.

- *Stereospecific:* In this case, the enzyme will only work with one stereoisomer. For example, D-glucose can serve as a substrate but L-glucose will not.

Inhibitors

An inhibitor is a molecule that interferes with the activity of an enzyme. For example, in competitive inhibition it does this by binding to the active site of the enzyme, preventing the natural substrate from doing so. This is shown in Fig. 9-6.

The fit of the inhibitor does not have to be exact; all it has to do is prevent the substrate from binding to the active site of the enzyme. Many poisons are inhibitors, as are many drugs. For example, an enzyme called *cytochrome oxidase* is involved in the transfer of electrons during respiration. *Cyanide* acts as a noncompetitive inhibitor with this enzyme, preventing it from functioning properly. As a result respiration is not possible; this is why cyanide is a deadly poison.

A class of drugs used as antibiotics called *sulfonamides* act as competitive inhibitors in bacteria. Growth and metabolic activity in bacteria depends on a molecule called

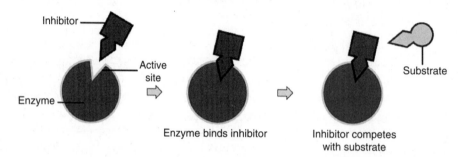

Figure 9-6　Other molecules can act like duplicate keys and fit the active site of an enzyme. This prevents the substrate from binding to the enzyme, and the reaction normally catalyzed by the enzyme cannot take place. (*Courtesy of Wikipedia.*)

p-aminobenzoic acid (PABA). This molecule is a substrate for an enzyme called dihydropteroate synthetase, which catalyzes the formation of tetrahydrofolic acid, which the bacteria need. Sulfonamides act as competitive inhibitors, competing with PABA for the active site on dihydropteroate synthetase, shutting down bacterial metabolism.

We can measure the efficiency by which an inhibitor interferes with a reaction. This is done by considering the degree of inhibition which is defined as

$$i = \frac{v_0 - v_i}{v_0}$$

Here, v_0 is the reaction rate when no inhibitor is present, while v_i is the inhibited reaction rate. Notice that if the inhibited reaction rate were 0 (that is the inhibitor completely shut off the reaction) then the degree of inhibition would attain its maximum value of 1. Obviously since inhibition will interfere with the reaction, the inhibited reaction rate must be less that the uninhibited reaction rate, that is, $v_i < v_0$ (notice that if they were equal, then the degree of inhibition would be zero). Hence the degree of inhibition ranges as $0 < v_i \leq 1$.

Inhibitors can act by competitive, noncompetitive, or uncompetitive inhibition in three ways:

- Competitive inhibition occurs if an increase in the concentration of the substrate is correlated with a decrease in the degree of inhibition. A competitive inhibitor has a similar chemical structure to the substrate allowing it to bind to the active site of the enzyme. This means that in competitive inhibition, the enzyme can bind to the substrate or the inhibitor. If the enzyme binds the substrate, it cannot bind the inhibitor, explaining why an increase in the concentration of the substrate is associated with a decrease in the degree of inhibition. Also note that increasing the amount of substrate can overcome the affect of the inhibitor and restore the increased rate of the catalyzed reaction.

- Noncompetitive inhibition occurs if the degree of inhibition is not affected by a change in the concentration of the substrate. In this case, the inhibitor can bind to the enzyme, or to the enzyme-substrate complex. The inhibitor does not bind to the same site as the substrate does, so the substrate can still bind to the enzyme in the presence of the inhibitor. This explains why the degree of inhibition is not affected by the concentration of the substrate in this case. A noncompetitive inhibitor affects the catalyzed rate of a reaction.

- Uncompetitive inhibition occurs when an increase in the concentration of the substrate results in an increase in the degree of inhibition.

COMPETITIVE INHIBITOR

- Causes K_m to increase. This means that the affinity of the enzyme for the substrate is decreased.
- Does not affect V_m, the maximum reaction rate because at high substrate concentration, the degree of inhibition is minimized. In fact at high substrate concentration the substrate competes out the inhibitor, and the inhibitor has no effect at all.
- Can bind to an alternate site besides the active site, causing a conformational change in the enzyme which prevents it from binding the substrate.

NONCOMPETITIVE INHIBITORS

- Have no affect on K_m.
- Bind to an alternative site (not the active site) on the enzyme, causing a conformational change in the enzyme.
- But they do not prevent substrate from binding. Instead, noncompetitive inhibitors inhibit the progress of the reaction to the products.

It is also possible for an inhibitor to be *mixed* with characteristics of competitive and noncompetitive inhibitors. An inhibitor is far more likely to be mixed than purely noncompetitive. Mixed inhibitors differ from purely noncompetitive inhibitors in the following way:

- For a mixed inhibitor, K_m is increased, thereby decreasing the affinity of the enzyme for the substrate. V_m is also affected.

Mixed inhibitors and purely noncompetitive inhibitors work best at high substrate concentrations, they decrease the maximum velocity V_m of the reaction.

This is because these types of inhibitors form an enzyme-substrate-inhibitor complex that prevents the formation of the product. The formation of this complex can take place in one of the following two ways:

$$E + I + S \rightarrow EI + S \rightarrow EIS$$

or

$$E + I + S \rightarrow ES + I \rightarrow EIS$$

An uncompetitive inhibitor cannot bind to the enzyme:

$$E + I \nrightarrow EI$$

In this case, the inhibitor can only bind to the enzyme-substrate complex:

$$E + I + S \rightarrow ES + I \rightarrow EIS$$

In the case of an uncompetitive inhibitor, the substrate itself is either involved in binding to the inhibitor or it causes a change in the structure of the enzyme allowing the inhibitor to bind. Uncompetitive inhibitors cause K_m to increase, resulting in an increased affinity of the enzyme for the substrate. However, since uncompetitive inhibitors work by binding to enzyme-substrate complex, when concentrations of the substrate are high the inhibitor works better. Hence it tends to decrease the maximum velocity V_m of the reaction.

Irreversible and Reversible Inhibition

Enzyme inhibition can be *irreversible* or *reversible*. If enzyme inhibition is irreversible, the inhibitor forms a strong bond (usually a covalent bond) with the enzyme, inactivating it. An irreversible reaction of the form $E + I \rightarrow EI$ takes place via a covalent bond, effectively permanently taking the enzyme out of circulation. Covalent bonds take longer to form so irreversible inhibition is a time-dependent process, with more enzyme being inactivated with increasing time. Reversible inhibition involves non-covalent bonding. Competitive, uncompetitive, and pure noncompetitive inhibitors are all examples of reversible inhibition.

Regulation of Enzyme Activity

Enzyme activity can be regulated in several ways. First, at the level of the chromosome, is the regulation of the expression of the genes that code for the manufacture of enzymes. In this way the amount of enzyme in the cell can be controlled. This is a slow process that acts over the course of hours, days, and weeks. In addition, enzymes in the cell can be broken down by proteolytic degradation, another slow process.

The activity of some enzymes is regulated by the presence of other molecules. An *allosteric site* is a site on an enzyme other than the active site where molecules can bind. In *allosteric regulation*, a molecule called an *effector* binds to the allosteric site. If the effector increases the activity of the enzyme, it is called an *allosteric activator*. On the other hand, if the effector decreases the activity of the enzyme, it is called an *allosteric inhibitor*. When the effector binds to the enzyme, it can alter the shape of the active site either enhancing or decreasing the affinity of the enzyme for the substrate. Allosteric enzymes tend to be located at branch points in metabolic pathways, so activating or inhibiting an allosteric enzyme can enhance or retard a given metabolic process. Regulatory enzymes control the rates of biochemical reactions that involve multiple steps. Good examples include glycolysis and the TCA cycle.

Another method that can be used by cells to regulate enzyme activity is by simply isolating them or locking them up. That is, the activity of the enzyme is regulated by limiting its access to substrate molecules. An example of this is the *lysosome,* a structure that restricts the activities of hydrolases by compartmentalizing them in a membrane structure until they are needed. When they must be used, the lysosome fuses with a vacuole containing the offending matter (an ingested bacteria or old organelle perhaps) that must be digested. The enzymes are dumped into the vacuole where they can gain access to the substrates they digest and go to work. Normally, the lysosome keeps these digestive enzymes tucked away so that they do not interfere with the normal operation of the cell by digesting healthy tissues.

Coenzymes and Cofactors

If an enzyme is in an inactive form, we call it an *apoenzyme.* In this case, the enzyme requires the presence of another compound called a *cofactor* in order to become active. When the enzyme is in the active form, we call it a *holoenzyme.* That is:

$$\text{Apoenzyme} + \text{Cofactor} \rightarrow \text{Holoenzyme}$$

The cofactor is not a protein, but may be some type of inorganic molecule such as a metal ion. Examples include manganese and copper. It is necessary for the cofactor to bind to the enzyme in order for the catalytic reaction mediated by the enzyme to occur.

When the cofactor is an organic molecule, we say that it is a *coenzyme.* The function of a coenzyme is to accept or donate chemical groups. Unlike an enzyme, a coenzyme is not a protein molecule. Many coenzymes are vitamins or derive from vitamins through metabolic processes (e.g., the B vitamins).

If a coenzyme becomes tightly bound to the enzyme, we say it is a *prosthetic group.* If it is loosely bound to the enzyme, we say it is a *cosubstrate.* Let's briefly review some important coenzymes. This list is very brief; it serves only to illustrate the types of roles that coenzymes play in important bioprocesses.

NAD$^+$

A key player in the electron transport chain in mitochondria and in all energy-liberating processes within cells is NAD$^+$ or *nicotinamide adenine dinucleotide.* NAD$^+$ is derived from vitamin B$_3$ (niacin) in the cell as follows:

$$\text{niacin} + \text{ribose} + \text{ADP} \rightarrow \text{NAD}^+$$

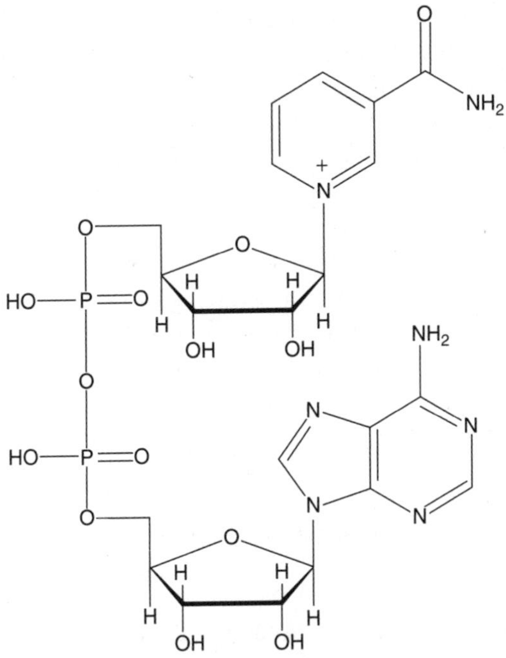

Figure 9-7 The NAD$^+$ molecule.

NAD$^+$ functions as an enzyme cofactor. The electron transport reaction in which a pair of electrons is accepted yielding NADH can be written as:

$$NAD^+ + H^+ + 2e^- \rightarrow NADH$$

In essence, the cofactor NAD$^+$ has accepted the equivalent of two hydride ions H$^-$. This reaction takes place in glycolysis, for example. NAD$^+$ is shown in Fig. 9-7.

FAD

The so-called *flavin nucleotides* are derived from the B vitamins and play a role in electron transport. In particular, flavin adenine dinucleotide (FAD) is synthesized from vitamin B$_2$ (riboflavin) and it functions as a prosthetic group for many enzymes where it plays its role in electron transport. In essence, FAD transports electrons by taking on two complete hydrogen atoms in the following reaction:

$$FAD + 2H^+ + 2e^- \rightarrow FADH_2$$

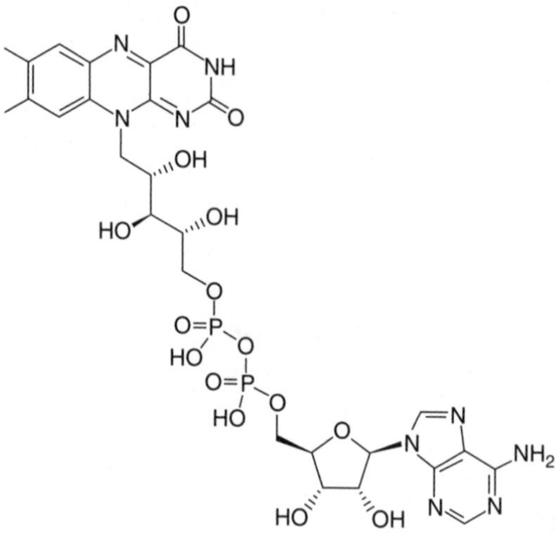

Figure 9-8   FAD is a coenzyme that covalently bonds to a dehydrogenase.

The type of enzyme which uses FAD as a coenzyme is called a *dehydrogenase*. FAD becomes linked to a dehydrogenase via a covalent bond. FAD is illustrated in Fig. 9-8.

It functions by playing a role in electron transport.

COENZYME A

Coenzyme A plays a role in the oxidation of pyruvate in the citric acid cycle and in the metabolism of fatty acids. When coenzyme A is bound to acetic acid, we call it *acetyl-coenzyme A* (Fig. 9-9). This molecule is very important in many biochemical reactions. Acetyl-coenzyme A or acetyl Co-A is a compound composed of adenosine diphosphate (ADP), pantothenic acid, and β-mercaptoethylamine. The chemical

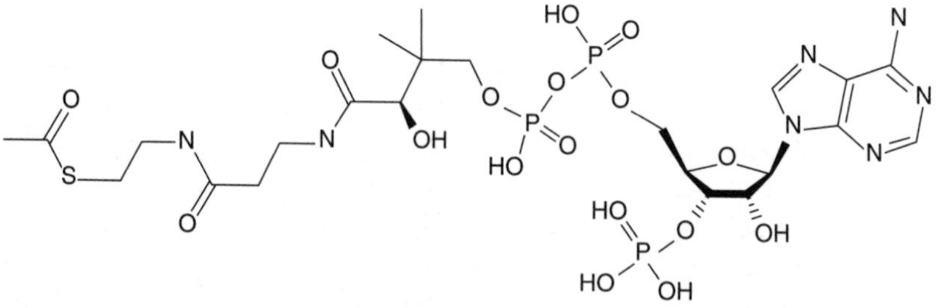

Figure 9-9   Acetyl-coenzyme A. (*Courtesy of Wikipedia.*)

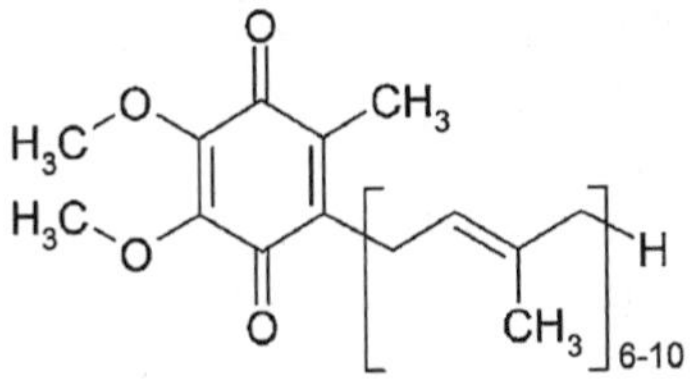

Figure 9-10   The chemical structure of coenzyme Q10. (*Courtesy of Wikipedia.*)

role of Co-A is as a carrier of acyl groups. This coenzyme plays a fundamental role in the body, the condensation of oxaloacetate with acetyl-CoA forming citric acid is the first step in the citric acid cycle.

COENZYME Q10

An important coenzyme used by the body is *coenzyme* Q10, which functions as a cofactor in the electron transport chain in the mitochondria. Therefore coenzyme Q10 is an important molecule in the generation of energy for the cell. Since it is involved in electron transport, it also functions as an antioxidant. It is found throughout the body, with the highest levels being in the heart, liver, kidneys, and the pancreas. The level of coenzyme Q10 in the body decreases with increasing age, moreover some cancer patients have abnormally low levels of coenzyme Q10. This may be because coenzyme Q10 may be involved in the maintenance of a healthy immune system. It may even be involved in cancer prevention; drugs with a chemical structure that is similar to coenzyme Q10 have been shown to kill cancer cells.

Coenzyme Q is a benzoquinone derivative. Coenzyme Q10 has 10 isoprene subunits. The number of isoprene subunits is used to classify the coenzyme Q; the mitochondria in mammalian cells have coenzyme Q10. The chemical structure of coenzyme Q10 is shown below in Fig. 9-10.

Isozymes

When there are two or more different enzymes that have a similar chemical structure and have in general similar taste for substrate molecules we call the group of enzymes *isozymes*. In a nutshell isozymes are different enzymes that catalyze the same chemical reactions. While they have similar chemical structures, they can behave very differently in the biochemical environment. This is because by changing a few amino acids in the enzyme we can radically alter its enzyme kinetics. So, while two isozymes might have a taste for the same substrate, the strength of that taste may be very different. In other words, the two isozymes will have different K_m values or different affinities for the same substrate. A common example is the two enzymes hexokinase and glucokinase. We have already seen these two enzymes many times; hexokinase has a large affinity for glucose while glucokinase does not. The reason there are two enzymes is because they act in different locations of the body for different purposes. Glucokinase is found primarily in the liver, where it is involved in the synthesis of glycogen (a stored form of carbohydrates). Hexokinase is found primarily in the skeletal muscle, where it can be utilized to extract energy from glucose immediately.

If an enzyme requires a cofactor/coenzyme, different isozymes will require different cofactors/coenzymes.

Summary

An enzyme is a protein molecule which acts to dramatically increase the rate at which a chemical reaction will occur. Generally speaking, an enzyme binds to a substrate (at the active site) forming an enzyme-substrate complex. The reaction occurs and the desired product is liberated. If the substrate immediately fits the active site of the enzyme, we say that the formation of the enzyme-substrate complex works using the lock and key model. If the substrate induces a small change in the conformation of the enzyme so that they can bind tightly, we call it the induced fit model.

An enzyme works by lowering the activation energy for a chemical reaction. Enzyme kinetics, which quantitatively describes how an enzyme speeds up a reaction, is characterized by the Michaelis-Menten constant K_m. If K_m is small, this means that an enzyme has a high affinity for a given substrate. If K_m is large, the enzyme does not have a high affinity for the substrate.

An enzyme with a high turnover number can catalyze more reactions in a given amount of time. If an enzyme has a low K_m and a high turnover number, it will be a highly efficient enzyme.

Sometimes an enzyme requires the presence of a helper molecule in order to function. A molecule that helps an enzyme become active is called a cofactor (for an inorganic molecule) or a coenzyme (for an organic molecule).

Quiz

1. The effect of an enzyme is to
   (a) lower the free energy of the reactants.
   (b) lower the activation energy for a reaction.
   (c) minimize the difference in free energy of the products and reactants.
   (d) raise the activation energy for a reaction.

2. Which of the following is not a true statement about enzymes?

  (a) Enzyme activity is enhanced by rising temperatures, to a point.

  (b) Very high temperatures tend to denature enzymes.

  (c) Enzymes typically have a range of pH over which they are effective.

  (d) Enzyme activity is enhanced by rising pH.

  (e) Enzyme activity is influenced by temperature, pH, and the concentration of substrate.

3. A noncompetitive enzyme inhibitor differs from a competitive enzyme inhibitor in which way?

  (a) There are no significant differences.

  (b) A noncompetitive inhibitor works well at low and high concentrations of substrate. A competitive inhibitor does not work well at low concentrations of substrate.

  (c) A noncompetitive inhibitor works well at low and high concentrations of substrate. A competitive inhibitor does not work well at high concentrations of substrate.

  (d) A competitive inhibitor causes K_m to decrease, while a noncompetitive inhibitor causes K_m to increase.

4. The Michaelis-Menten constant K_m

  (a) describes the affinity of an enzyme for a substrate. If K_m is small, then the enzyme has little affinity for the substrate.

  (b) describes the affinity of an enzyme for a substrate. If K_m is small, then the enzyme has a great deal of affinity for the substrate.

  (c) describes the maximum reaction rate for an enzyme.

  (d) describes the affinity of an enzyme for a substrate. If K_m is large, then the enzyme has a great deal of affinity for the substrate.

5. If K_m is large, which of the following reactions is favored?

  (a) $ES \rightarrow E + S$

  (b) $E + S \rightarrow ES$

  (c) $ES + I \rightarrow EI + S$

  (d) $EI + S \rightarrow ES + I$

6. In the formation of an enzyme substrate complex, the reaction rate

  (a) tends to the maximum V_m when the substrate concentration $[S] < K_m$.

  (b) depends solely on the Michaelis constant K_m.

(c) is highly dependent on the substrate concentration $[S]$, with $V \to \frac{1}{2}V_m$ if $[S] = K_m$.

(d) goes to $V \approx V_m K_m$ when the substrate concentration is very small.

7. A mixed inhibitor differs from a purely noncompetitive inhibitor in that

(a) a mixed inhibitor always binds to the active site.

(b) a mixed inhibitor works best at low substrate concentrations.

(c) K_m is increased in the case of a mixed inhibitor.

(d) K_m is unchanged for a mixed inhibitor, but K_m increases for a noncompetitive inhibitor.

8. Irreversible inhibition

(a) involves a strong covalent bond between the enzyme and inhibitor.

(b) involves the transfer of a proton.

(c) is usually effected using van der Waals interactions.

(d) inhibition is always reversible.

9. Competitive inhibitors differ from noncompetitive inhibitors in that

(a) a competitive inhibitor causes K_m to increase while a noncompetitive inhibitor does not effect K_m.

(b) a noncompetitive inhibitor causes K_m to increase while a competitive inhibitor does not effect K_m.

(c) a noncompetitive inhibitor only binds to a site in the enzyme causing a conformational change while a competitive inhibitor can only bind to the active site of the enzyme.

(d) a competitive inhibitor acts to decrease V_m, while a noncompetitive inhibitor has no effect.

10. Two isozymes

(a) have the same amino acid sequences but use different coenzymes.

(b) have slightly different amino acid sequences, but can have radically different enzyme kinetics.

(c) are different enzymes but have the same affinities for the same substrates.

(d) are two enzymes coded for by different DNA sequences.

CHAPTER 10

Glycolysis

Life cannot exist without taking in energy from the surrounding environment and transforming it for useful purposes. In this chapter, we see how the metabolism of carbohydrates is useful in the life process. We begin here with an examination of *glycolysis,* which is the most basic of reactions used to liberate energy in the cell from a single glucose molecule under anaerobic conditions. Glycolysis is a process that takes place in the cytoplasm. To understand glycolysis you will need to know

- The energy input and output of glycolysis.
- The steps in glycolysis.
- The enzymes used and intermediate products in glycolysis.
- About pyruvate, the end product of glycolysis.
- Lactate and ethanol fermentation.
- The Cori cycle.

A few general notes about energy reactions are good to keep in mind

- If $\Delta G^{\circ\prime} < 0$, then the reaction is exergonic and therefore spontaneous.
- A step that has a large $\Delta G^{\circ\prime}$ which is negative tends to be a regulated step.

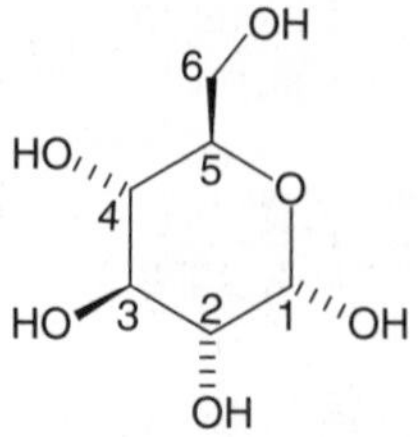

Figure 10-1 A glucose molecule has 6 carbon atoms. In glycolysis, it will be cleaved into two 3 carbon atom molecules.

- If a cell has a high adenosine triphosphate (ATP) level, it is in a *high energy* state.
- If a cell has a low ATP level [and conversely a high adenosine diphosphate (ADP) level] it is in a *low energy* state.

Introduction

Glycolysis, which literally means *splitting of sugar*, is a process whereby a glucose molecule is split in two in order to generate ATP. This is the simplest form of metabolism in the cell, which takes place in the absence of oxygen and is utilized under at least some conditions by all types of cells. Moreover, glycolysis converts one molecule of glucose into 2 molecules of pyruvate which can then be used to generate more ATP in aerobic organisms.

A glucose molecule, shown in Fig. 10-1 to refresh your memory, contains 6 carbon atoms; a molecule of pyruvate, shown in Fig. 10-2, has 3 carbon atoms.

The release of energy from a glucose molecule using glycolysis is probably of very ancient origin. We know this because

- It is an anaerobic process, that is, no oxygen is required.
- It occurs even in very simple cells.
- After glycolysis more "advanced" aerobic organisms use oxygen to generate additional energy. Pyruvate can be oxidized in the citric acid cycle, which we discuss in the following chapter.

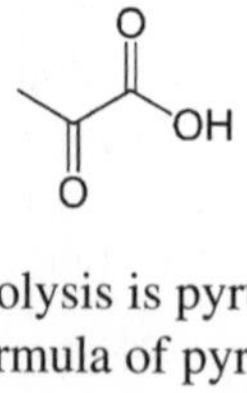

Figure 10-2 The end product of glycolysis is pyruvate, a molecule with 3 carbon atoms. The chemical formula of pyruvate is $C_3H_4O_3$.

Glycolysis occurs in the cytoplasm and it requires a total of 10 enzymes along with the investment of 2 ATP molecules. These 2 ATP molecules are used in the following ways:

- An ATP molecule donates a phosphate group to glucose giving *glucose phosphate.*

- The second ATP molecule donates a phosphate turning fructose-6-phosphate to fructose 1,6-bisphosphate.

The reasons for these ATP investments will become clear below. The overall reaction, which requires the production of 9 intermediate molecules, is:

$$C_6H_{12}O_6 + 2ADP + 2NAD^+ + 2P_i \rightarrow 2C_3H_4O_3 + 2ATP + 2NADH + 2H^+ + 2H_2O$$

Note that,

- The metabolism of glucose into two molecules of pyruvate yields a net gain of 2 ATP molecules. We invest 2 ATP molecules up front, and 4 ATP molecules are liberated during the cycle. This net gain in ATP provides a vital energy source that can be utilized in cellular processes.

- Glycolysis yields 2 NADH molecules and 2 protons. In the back of your mind, make a note that the cell requires NAD^+ to generate more ATP.

- In cells that can use aerobic metabolism, pyruvate can be used to synthesize more ATP in the citric acid cycle.

Glycolysis occurs in two phases we call the *preparatory phase* and the *pay off phase.* In the preparatory phase an investment of ATP molecules is required. In the pay off phase energy is liberated.

The Preparatory Phase

The preparatory phase of glycolysis requires five steps that prepare the glucose molecule so that it can liberate energy in the form of new ATP molecules. We will consider each step in turn. Remember that glycolysis takes place within the cytoplasm. The important input and output of the preparatory phase are

- *Input:* One glucose, 2 ATP, 5 enzymes.

- *Output:* Two molecules of glyceraldehyde 3-phosphate.

STEP 1: THE HEXOKINASE REACTION

The first step in glycolysis is the conversion of glucose into *glucose 6-phosphate*. This step is a *trapping reaction* that traps a glucose molecule in the cell, since glucose 6-phosphate cannot diffuse out of the cell. This produces a concentration gradient, since the concentration of glucose inside the cell is lower than it is outside, which continues to encourage glucose to enter the cell. In contrast to the transport of glucose across the cell membrane, there is no mechanism of transport of phosphorylated sugars across the cell membrane. As a result glucose 6-phosphate stays put inside the cell.

The hexokinase-mediated step in glycolysis requires 1 ATP molecule to proceed. It is catalyzed by the enzyme *glucose 6-phosphate isomerase* which is usually called *hexokinase*. This reaction is illustrated in Fig. 10-3.

STEP 2: THE FORMATION OF FRUCTOSE 6-PHOSPHATE

The next step in the preparatory phase of glycolysis is the formation of *fructose 6-phosphate*. This is a freely reversible reaction, which is needed in order to make C-3 suitable for cleavage of the molecule in two. This is illustrated in Fig. 10-4.

STEP 3: FORMATION OF FRUCTOSE 1,6-DIPHOSPHATE

The third step in the preparatory phase requires the investment of another ATP molecule and is catalyzed by the enzyme *phosphofructokinase* (PFK). This step adds another phosphate group to the molecule, transforming fructose 6-phosphate

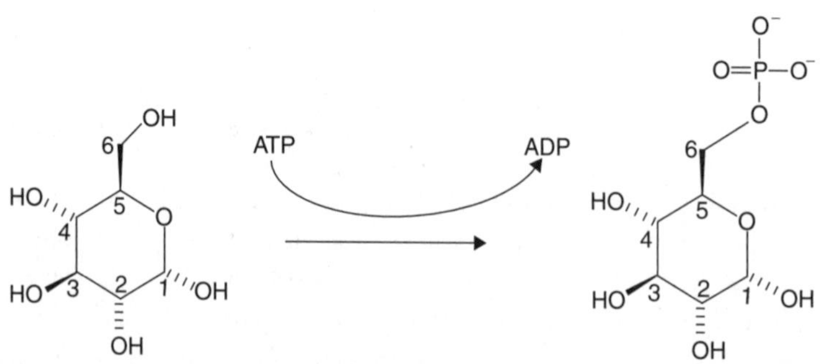

Figure 10-3   By adding a phosphate group to a glucose molecule, we can inhibit its diffusion out of the cell. This is the first step in glycolysis, and it costs 1 ATP molecule.

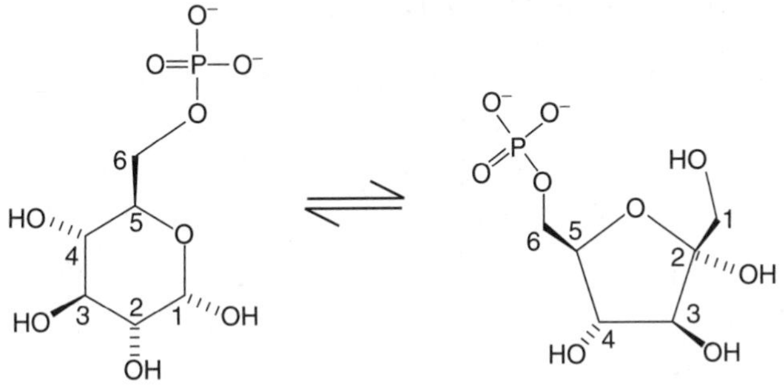

Figure 10-4   In step two of glycolysis, glucose 6-phosphate undergoes an isomerization reaction to yield fructose 6-phosphate, which activates C-3 for cleavage of the molecule.

into fructose 1,6-bisphosphate. This step has a large and negative $\Delta G^{\circ\prime}$ and so is a very regulated step.

STEP 4: CLEAVAGE OF FRUCTOSE 1,6-BIPHOSPHATE INTO GLYCERALDEHYDE 3-PHOSPHATE AND DIHYDROXYACETONE

The next step in the preparatory phase is the actual cleavage of the molecule, as illustrated in Fig. 10-5. *Glyceraldehyde 3-phosphate* (GAPD) is the desired end product of the preparatory phase. To get two copies of the molecule, we need to convert *dihydroxyacetone* (DHAP) in one more step.

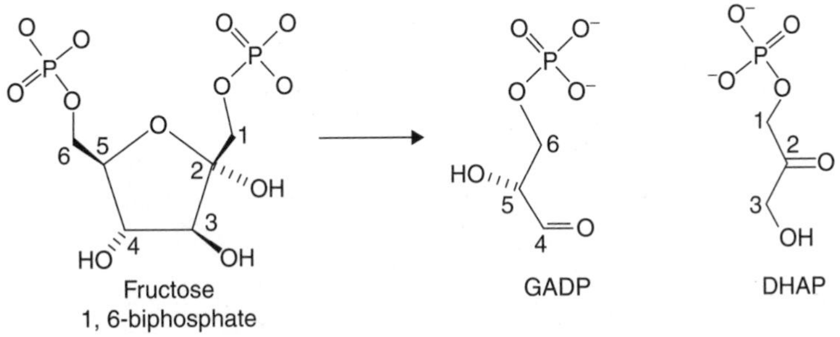

Figure 10-5   The cleavage step transforms fructose 1,6-biphosphate into GADP (the end product of the preparatory phase) and DHAP. One more step is required to convert DHAP into another copy of GADP.

STEP 5: CONVERSION OF DIHYDROXYACETONE INTO GLYCERADLEHYDE 3-PHOSPHATE

In the presence of an enzyme called *triose phosphate isomerase,* dihydroxyacetone is transformed into glyceraldehydes 3-phosphate, giving us two copies of the molecule that can be used as input to the pay-off phase.

The preparatory phase requires energy input and several enzymes. These are summarized in Table 10-1.

The preparatory phase of glycolysis can be summarized as follows:

- A glucose molecule (a 6-carbon compound) is converted into 2 molecules of glyceride 3-phosphate molecules (a 3-carbon compound).

- Two molecules of ATP must be invested to provide the energy required to drive the reaction at steps 1 and 3.

- In the pay-off phase, *each* molecule of glyceraldehyde 3-phosphate will lead to the formation of 2 molecules of ATP. A total of 4 ATP molecules are therefore produced for each glucose. This gives a net gain of 2 ATP molecules.

The Pay-Off Phase of Glycolysis

In the pay-off phase of glycolysis, we see a net gain of 2 ATP molecules, and hence a net gain in energy for the cell. The pay-off phase is characterized by the following input and output:

- *Input:* Two glyceraldehydes 3-phosphate molecules from the preparatory phase of the reaction. In addition, the required input includes 2 NAD$^+$ molecules, 2 P$_i$ and 5 enzymes.

- *Output:* Two Pyruvate molecules and 4 ATP molecules.

Table 10-1   Summary of Energy Requirements and Enzymes Used in Each Step of the Preparatory Phase of Glycolysis

Step	Energy Input	Required Enzyme or Comments
1	1 ATP	Hexokinase. Keeps glucose in the cell.
2	—	Freely reversible reaction. Catalyzed by phosphoglucose isomerase.
3	1 ATP	Enzyme for this step is phosphofructokinase. Prepares molecule for cleavage.
4	—	Fructose 1,6-biphosphate aldolase. Cleavage step, reversible reaction.
5	—	Triose phosphate isomerase.

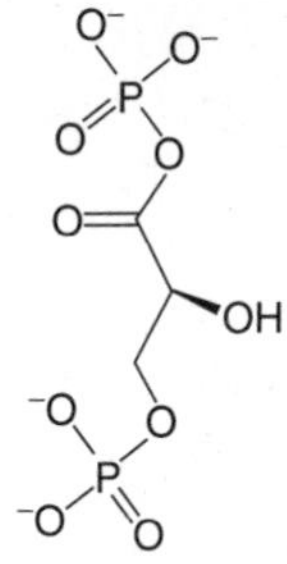

Figure 10-6   1,3-bisphosphoglycerate has two energy rich phosphate groups.

The steps in the pay-off phase are as follows, where we designate the steps 6–10 to follow step 5 of the preparatory phase.

STEP 6: OXIDATION OF GLYCERALDEHYDE 3-PHOSPHATE

Each molecule of glyceraldehyde 3-phosphate is oxidized producing 1,3-bisphosphoglycerate (see Fig. 10-6). This requires the input of $NAD^+ + P_i$ and liberates an NADH molecule and a proton (H^+).

STEP 7: SUBSRATE LEVEL PHOSPHORYLATION

In the next step, 1,3-bisphoglycerate is transformed into 3-phosphoglycerate (see Fig. 10-7). This is an *energy-yielding step* which results in the production of 1 ATP molecule for each 1,3-bisphosphoglycerate, and so a total of 2 molecules of ATP are produced for each initial glucose molecule.

STEP 8: MUTASE REACTION

Phosphoglycerate mutase catalyzes the transfer of the phosphate group from C-3 of 3-phosphoglycerate to C-2. So, each molecule of 3-phosphoglycerate is transformed into 2-phosphoglycerate (see Fig. 10-8).

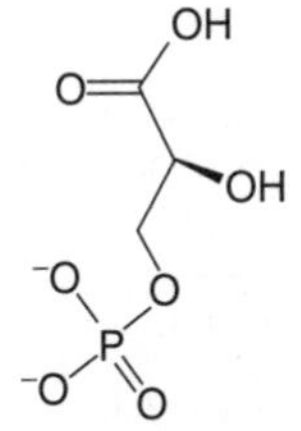

Figure 10-7   After phosphorylation, after a gain of an ATP we are left with 3-phosphoglycerate, which has one more phosphate group available.

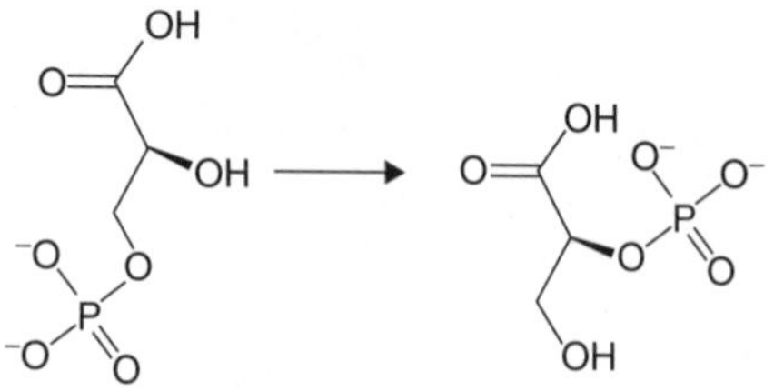

Figure 10-8   In step 8, the phosphate group is moved into a position that will be more favorable for energy liberation. Starting with 3-phosphoglycerate, we end up with 2-phosphoglycerate on the right.

STEP 9: ENOLASE-MEDIATED STEP

Each 2-phosphoglycerate molecule undergoes a dehydration reaction liberating 1 water molecule. The purpose of this step is to transform 2-phosphoglycerate into a form that is more suitable to release a higher amount of energy in the next step. To see why this is necessary, noting that the formation of 1 ATP molecule requires 30 kJ/mol and that the hydrolysis of 2-phosphoglycerate has a $\Delta G^{\circ\prime} = -16$ KJ/mol, we see that we are short of the energy required to synthesize an ATP. This step creates a compound, *phosphoenol-pyruvate* with enough punch to synthesize ATP (see Fig. 10-9). As indicated by the title given to this step, the enzyme which catalyzes it is called *enolase*.

STEP 10: SYNTHESIS OF PYRUVATE

In the final step of glycolysis, each phosphoenolpyruvate molecule is transformed into pyruvate. This is the final *energy-yielding* step in glycolysis, with the transformation of each phosphoenolpyruvate molecule into pyruvate yielding one ATP. Therefore a total of 2 ATP molecules are created in this step.

The pay-off phase of glycolysis is summarized in Table 10-2. The end result is that the preparation phase cost us 2 ATP molecules, and the pay-off phase generates 4 ATP molecules, for a net gain of 2 ATP molecules per input glucose. The reaction also costs us 2 molecules that will have to be replaced.

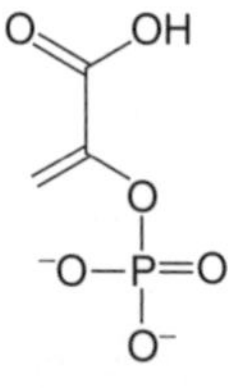

Figure 10-9   The last intermediate product in glycolysis, phosphoenolpyruvate. Phosphorylation of the molecule to convert ADP into ATP is now energetically favorable.

Table 10-2 A Summary of Steps Used in the Pay-Off Phase

Step	Enzyme Required	Products
6	Glyceraldehydes 3-phosphate dehydrogenase	NADH, H^+
7	Phosphoglycerate kinase	1 ATP
8	Phosphoglyceromutase	—
9	Enolase	$1 H_2O$
10	Pyruvate kinase	1 ATP

In the table, we consider 1 input glyceraldehydes 3-phosphate molecule, so the products should be doubled to get the total for the glycolysis of a single glucose molecule.

The pay-off phase of glycolysis can be summarized as follows:

- It requires the input of 2 glyceraldehyde 3-phosphate molecules.
- Each input molecule results in the output of 1 NADH, 1 H^+, and 1 H_2O.
- Each input molecule yields 2 ATP molecules, in steps 7 and 10 for a total yield of 4 ATP molecules.
- Since it costs 2 ATP molecules to set up the pay-off phase, the *net yield* is 2 ATP molecules.
- Glycolysis is an exergonic reaction. In particular, the conversion of glucose to 2 pyruvate molecules has $\Delta G^{\circ\prime} = -147$ kJ/mol.
- The conversion of ADP to ATP is endergonic and hence requires energy input. In this case $\Delta G^{\circ\prime} = +30$ kJ/mol.
- Therefore the net change in free energy is $\Delta G^{\circ\prime} = (-147 + 60)$ kJ/mol = -87 kJ/mol. Since $\Delta G^{\circ\prime} < 0$, the overall reaction is exergonic and is a favored reaction.
- Steps 1, 3, 7, and 10 in glycolysis are exergonic. Steps 2, 4, 5, 6, 8, and 9 are endergonic.

Glycolysis is summarized schematically in Fig. 10-10. We now review the regulatory steps in glycolysis.

Hexokinase Regulation

The first step in glycolysis requires the enzyme hexokinase to proceed. This is a rate-limiting step in the reaction because it is governed by a process called *product inhibition*. That is, the product in the first step, glucose 6-phosphate, actually inhibits

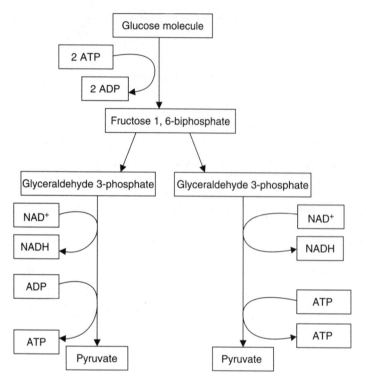

Figure 10-10 A schematic summary of glycolysis.

the activity of the enzyme that catalyzes the reaction, hexokinase! When hexokinase is not active, glucose enters the cell at will.

In mammals, glucose enters the cell from the bloodstream, where it is rapidly converted to glucose 6-phosphate by hexokinase. When the saturation point is reached for this reaction, glucose will diffuse out of the cell and back into the bloodstream. Hence hexokinase inhibition

- Prevents too much accumulation of glucose 6-phosphate in the cell. If the cell's energy requirements are already met and glucose 6-phosphate cannot be utilized, too much of it would accumulate and cause the cell to enlarge.

- When a cell's energy levels are met and production of glucose 6-phosphate is shut off, the extra glucose leaves the cell through diffusion.

- In liver cells, *glucokinase,* rather than hexokinase, is present. This enzyme also converts glucose into glucose 6-phosphate, but in the liver glucose is converted into glycogen and stored. This means that under normal conditions, excess glucose does not accumulate in cells but diffuses into the blood stream, where it then enters the liver. Glucokinase helps store this excess glucose as glycogen.

Phosphofructokinase Regulation

ADP and adenosine monophosphate (AMP) enhance the activity of phosphofructokinase (PFK) while ATP and NADH inhibit it. Let's remind ourselves that a cell can be found in a *low energy state* or a *high energy state* where

- A low energy state is one where concentrations of ADP and AMP are relatively high while concentrations of ATP are relatively low.

- A high energy state is one where concentrations of ADP and AMP are relatively low while concentrations of ATP are relatively high.

ATP exerts *allosteric control* over the PFK enzyme by binding to a regulatory site on the molecule. This slows the reaction that converts fructose 6-phosphate into fructose 1,6-biphosphate and hence slows the production of pyruvate when energy is not needed by the cell. The way this works is based on the fact that PFK has two conformational states that scientists have called the R and the T states. A high ATP concentration causes PFK to assume the T state which decreases its affinity for fructose 6-phosphate, shutting down glycolysis. On the other hand, a high concentration of AMP and ADP causes PFK to assume the R state, increasing its affinity for fructose 6-phosphate, driving the glycolysis reaction.

Pyruvate Kinase Regulation

The final regulating step in glycolysis is the final step in the reaction, which involves the production of pyruvate. This step is catalyzed by the enzyme *pyruvate kinase* which is activated by the presence of fructose 1,6-biphosphate. Pyruvate kinase is inhibited by a high ATP concentration.

Hence, when a cell is in a high-energy state and therefore has a high concentration of ATP, the production of pyruvate will be reduced. In contrast, a low energy state, in which there is a low concentration of ATP, will bring up levels of pyruvate and ATP because pyruvate kinase will be active. The effect of a low energy state of the cell is summarized in Fig. 10-11.

THE FATE OF PYRUVATE

Glycolysis, like many things of old, is an inefficient way to get things done. By itself, it does not completely exploit the energy in a glucose molecule. In fact, the complete oxidation of glucose would liberate 38 ATP molecules:

$$C_6H_{12}O_6 + 6\,O_2 \rightarrow 6\,CO_2 + 6H_2O$$

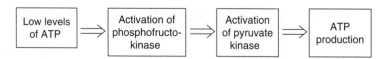

Figure 10-11 The effect of low levels of ATP is to increase enzyme activity for enzymes used in glycolysis, increasing ATP production.

This shows that with oxygen around, there must be a better way to extract energy from glucose. It turns out that there are better ways, but nature had to wait until there was oxygen around before it could find them. And typically, evolution is conservatively making use of what's already around and building on it to create new designs. For example, your brain is a layered structure that has within it the older less developed brains of a less sophisticated mammal (a rabbit say) and a reptile (the so-called R-complex). Evolution kept what worked and slapped on a cerebral cortex on top of it.

In much the same way, biochemically speaking, organisms had a process that was able to liberate *some* ATP, which kept life going even if it didn't get out all the energy that glucose had to offer. In fact, for the complete oxidation of glucose $\Delta G^{\circ\prime} = -2870$ kJ/mol, which means that glycolysis only liberates 3% of the energy that is available in a glucose molecule. For life to move forward there had to be a better way. This was done by keeping glycolysis around and getting the rest of the energy out using a process called *oxidative phosphorylation.* A comparison of anaerobic and complete aerobic respiration of glucose is shown in Table 10-3.

However, before we get to that, which is the topic of the next chapter, we have another problem on our hands. Glycolysis requires the input of NAD$^+$. As far as we've seen, there isn't any way to make more of it. So the first item of business is to find a way to make more NAD$^+$. This would be important for a primitive anaerobic organism or for a mammal when oxygen was not immediately available. For glycolysis to continue, and hence for the production of ATP to continue, the cell needs a continuous supply of NAD$^+$. Otherwise it will run out, glycolysis will not be possible, ATP will run out and the cell will die.

Table 10-3 A Comparison of the Net Gains for Glycolysis Alone and Oxidative Phosphorylation

	Anaerobic (Glycolysis Alone)	**Aerobic (Oxidative Phosphorylation)**
ATP Production	2	38
$\Delta G^{\circ\prime}$	−87 kJ/mol	−2870 kJ/mol

It turns out that there are three possible fates for the pyruvate molecule:

- Conversion into lactate
- Anaerobic fermentation converting pyruvate into ethanol and carbon dioxide
- Oxidative phosphorylation, where pyruvate enters the citric acid cycle and liberates the maximum amount of ATP possible

The last item is the subject of the next chapter. Here we consider lactate and ethanol fermentation.

Lactate

The fate of pyruvate depends on whether or not the cell is in the presence of oxygen and what type of cell it is. So in many cases, pyruvate cannot be oxidized, whether it's due to the fact that the cell in question is primitive or if there is no oxygen around. While not ideal, this is not a problem as far as the production of ATP is concerned since the glucose $\rightarrow$ pyruvate pathway has a net yield of 2 ATP molecules. The problem is that glycolysis reduces NAD^+ to NADH. This is the last we see of NAD^+ in the reaction, and if there isn't a way to produce more of it glycolysis and hence energy production will stop. The key to continuing glycolysis is to reoxidize NADH to NAD^+. So the cell can convert pyruvate into lactate. The purpose of this reaction is to regenerate the supply of NAD^+ so that the production of ATP can continue. This process is called *lactate fermentation.*

In the red blood cells and the skeletal muscle cells, this is done by converting pyruvate to lactate (Fig. 10-12). This reaction is catalyzed by an enzyme called *lactate dehydrogenase.*

There are five forms of the lactate dehydrogenase enzyme called *isozymes.* Which isozyme or form of lactate dehydrogenase the cell contains determines how it processes pyruvate. These are designated as M (for muscle) and H (for heart) as M4, M3H, M2H2, MH3, and H4. At one extreme, skeletal muscle cells contain a predominance of M4, while at the other extreme, heart muscle cells contain a predominance of H4. M4 and H4 can be characterized by their affinity for pyruvate and their turnover rate, with the other isoenzymes falling somewhere in between depending how much M they have and how much H they have. In short,

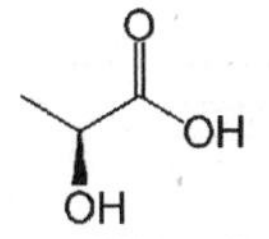

Figure 10-12   A lactate molecule.

- H4 has a high affinity for pyruvate, but a *low* turnover number. This means that if concentrations of pyruvate are high, H4 converts pyruvate to lactate at a slower rate than M4.
- M4 has a low affinity for pyruvate, but a *high* turnover number. This means that M4 converts pyruvate to lactate at a higher rate than H4.

The enzymes and where they can be found are as follows:

- H4: Heart muscle cells have a predominance of H4. Heart cells tend to oxidize pyruvate to CO_2.
- MH3: Found in monocytes and macrophages.
- M2H2: Found in the lungs.
- M3H: Found in the kidneys.
- M4: Found in the liver and skeletal muscle.

During periods of exercise, skeletal muscle gets the lion's share of its ATP from glycolysis, and so it needs a continuous supply of NAD^+. The process of using lactate to regenerate the supply of NAD^+ under anaerobic conditions is part of a feedback loop called the *Cori cycle,* a process discovered by Carl and Gerty Cori who won the Nobel prize in 1937. Lactate produced by skeletal muscles can be recycled back into glucose using a feedback loop with the liver. In the initial stage of the Cori cycle, which is illustrated in Fig. 10-13, glucose is converted into pyruvate and then to lactate with the help of the enzyme lactate dehydrogenase. The lactate is then released into the blood stream.

To summarize, under anaerobic conditions or exercise, skeletal muscle uses glycolysis to obtain energy. To replenish the supply of NAD^+, lactate is produced, which then diffuses into the bloodstream. Once there it can be taken up by the liver, where it can be processed in the second phase of the Cori cycle.

It turns out that lactate dehydrogenase not only converts pyruvate to lactate, it also can convert lactate *back* to pyruvate. Then, through a series of steps to reverse glycolysis, it can be transformed into glucose 6-phosphate. This process is called *gluconeogenesis.* To complete the cycle, the liver transforms the glucose 6-

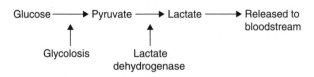

Figure 10-13　In the first phase of the Cori cycle, lactate produced during glycolysis in the skeletal muscle is released into the blood stream.

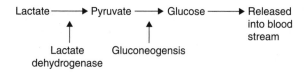

Figure 10-14   In the liver, the Cori cycle is completed. Lactate is converted back into glucose, which diffuses into the bloodstream where it can be taken up by skeletal muscle, and used to make more ATP.

phosphate back into glucose. It can then diffuse back into the blood stream where it can be taken up by the skeletal muscle, starting a new phase of the Cori cycle, as illustrated in Fig. 10-14. This completes the Cori cycle.

The Cori cycle requires 6 ATP in the gluconeogenesis phase which takes place in the liver. Since there is a net gain of 2 ATP molecules during glycolysis in the skeletal muscle, there is a net *loss* of 4 ATP molecules in the Cori cycle. The function of the Cori cycle is to provide a temporary, readily available supply of glucose to the skeletal muscles during exercise.

More on the Liver

Liver cells utilize an enzyme called *glucokinase* that is used in place of hexokinase during glycolysis. While hexokinase has a $K_m = 0.1$ mM for glucose, for glucokinase $K_m = 10$ mM. A high K_m implies a low affinity. Under normal conditions, the concentration of glucose in the bloodstream is 5 mmol/L. Two points to keep in mind are

- Cells that utilize hexokinase rapidly process glucose, converting it into glucose 6-phosphate.
- Liver cells have a tendency to *deliver* glucose to the bloodstream.

When the blood glucose level is high, glucokinase is the most active. This allows the liver to trap circulating glucose. It can be converted into glycogen and stored. On the other hand, the relatively low K_m for hexokinase means that a cell which utilizes this enzyme (a skeletal muscle cell or neuron say) will continue to use glucose even when the blood glucose level is low.

When the blood glucose level is low, there is not much glucokinase activity. In contrast, hexokinase activity is just about constant for any blood glucose level.

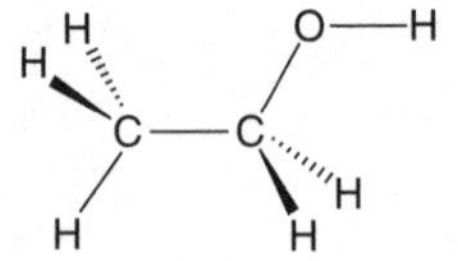

Figure 10-15 Yeast need to regenerate their supply of NAD$^+$ in anaerobic conditions. The resulting byproduct is ethanol.

Ethanol Fermentation

In more primitive organisms, such as in yeast, to the delight of many, there is another route for pyruvate after glycolysis called *ethanol fermentation*. This reaction, which regenerates a supply of NAD$^+$ for the yeast cell under anaerobic conditions, takes place in two steps:

- Pyruvate is converted to acetaldehyde.

- Acetaldehyde is converted to ethanol (ethanol is shown in Fig. 10-15) with the reduction of 1 NADH molecule liberating an NAD$^+$.

This reaction requires two enzymes:

- Pyruvate decarboxylase: This enzyme catalyzes the pyruvate → acetaldehyde reaction.

- Alcohol dehydrogenase: This enzyme catalyzes the acetaldehyde → ethanol reaction.

Ethanol fermentation differs from lactate fermentation in one fundamental way. In lactate fermentation, when oxygen becomes available lactate is oxidized back into pyruvate. However, pyruvate cannot be recovered in ethanol fermentation. Instead, a yeast cell converts acetaldehyde into acetic acid which is oxidized to carbon dioxide in the citric acid cycle.

The so-called *Pasteur effect* is useful in the production of alcoholic beverages. When living under anaerobic conditions, the amount of sugar consumed by yeast is significantly higher. To see why this is so, note that while the production of ATP using glycolysis is quite a bit lower than what it would be using oxidative phosphorylation, the *rate* of ATP production using glycolysis is higher. In fact, it can be up to 100 times faster than the rate of ATP production using oxidative phosphorylation.

Other Sugars in Glycolysis

Other sugars can be utilized in the glycolysis reaction for the generation of ATP. The sugar in question does not enter glycolysis at the initial stage; rather it enters the cycle at some later step. The step where glycolysis is initiated using other sugars

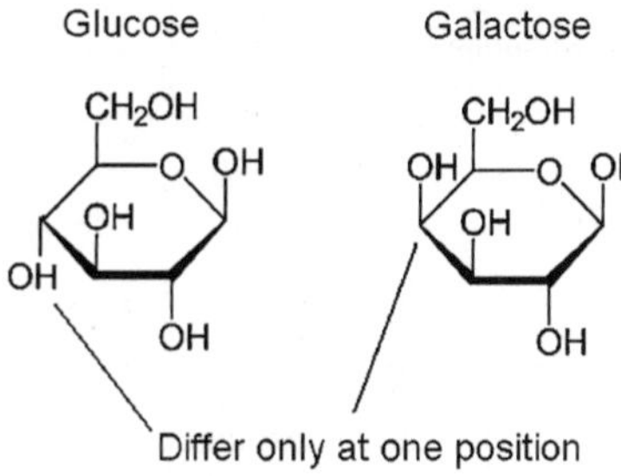

Glucose   Galactose

Differ only at one position

Figure 10-16   Glucose and galactose differ only at one position, but galactose must be converted into a glucose compound in order to enter glycolysis.

depends on what type of sugar is being utilized, and in some cases on what type of cell is being considered. Let's look at a few examples.

GALACTOSE

Galactose finds its way into the biochemist's body through the consumption of milk, or rather through the ingestion of lactose or milk sugar. While galactose is a monosaccharide, differing from glucose only at a single position (see Fig. 10-16), its entry into glycolysis is via a complicated path. An enzyme called *galactokinase* catalyzes the transformation of galactose into *galactose 1-phosphate,* which is then converted into a compound called *UDP-galactose. UDP-glucose-4-epimerase* then transforms this molecule into *UDP-glucose,* which can then be incorporated into glycogen.

Alternatively, instead of being incorporated into glycogen, *uridyl transferase* can catalyze the conversion of UDP-glucose into glucose 1-phosphate. Then, *phosphoglucomutase* catalyzes the transformation of this molecule into glucose 6-phosphate which can then be incorporated into glycolysis. This process is shown in Fig. 10-17.

FRUCTOSE

Sucrose, commonly known as *table sugar,* consists of a glucose and a fructose molecule bound together (see Fig. 10-18). An enzyme called *sucrase* cleaves the sucrose molecule into glucose and fructose. In muscle cells, fructose can be converted into fructose 6-phosphate which can then be utilized in glycolysis in step 3. In the liver, fructose does not enter glycolysis until a later step. Instead, it is converted into fructose 1-phosphate which then transforms into dihydroxyacetone phosphate + glyceraldehyde. The glyceraldehyde molecule can be transformed into glyceraldehyde 3-phosphate that can then be utilized in glycolysis at step 5.

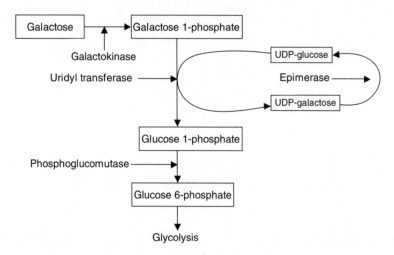

Figure 10-17   The complicated entry of galactose into glycolysis.

MANNOSE

Another monosaccharide that can be incorporated into glycolysis is *mannose* (see Fig. 10-19). This sugar comes from the digestion of polysaccharides and glycoproteins. Hexokinase phosphorylation converts mannose into *mannose 6-phosphate.* This molecule can be converted to fructose 6-phosphate by the enzyme *phosphomannose isomerase.* At this point, mannose can enter glycolysis at step 3.

Glycolysis and Cancer

Suprisingly, glycolysis is closely linked to cancer. In mammals, normal cells will not preferentially depend on glycolysis in the presence of oxygen because although it generates ATP at a higher rate than oxidative phosphorylation, it only liberates

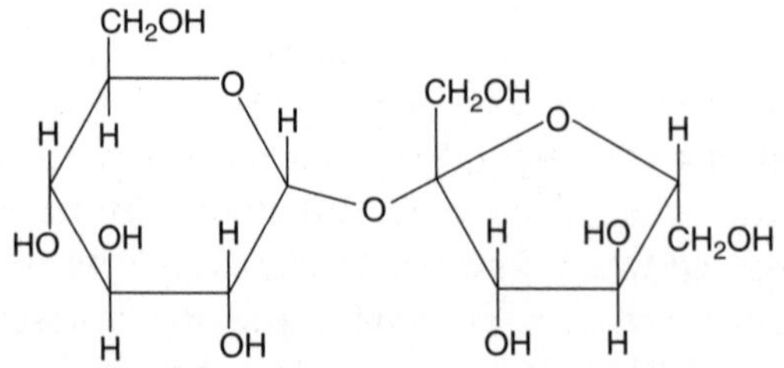

Figure 10-18   Sucrose can be cleaved into glucose + fructose by the enzyme sucrase.

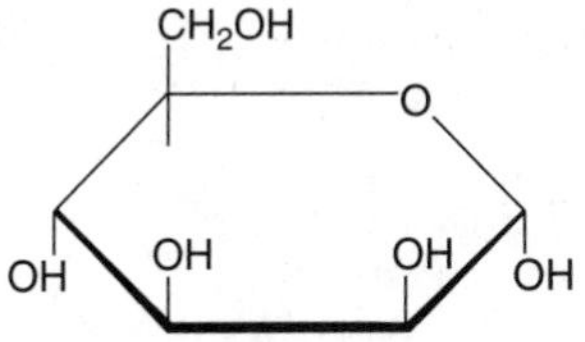

CH$_2$OH

O

OH   OH   OH   OH

Figure 10-19   Mannose.

3% of the energy available in a glucose molecule. So oxidation is the preferred method so that cells can extract as much energy as possible.

However, in the 1920s the German scientist Otto Warburg, who won the 1931 Nobel Prize in Physiology and Medicine for his discovery, realized that cancer cells utilize glycolysis even in the presence of adequate levels of oxygen. This process is called *aerobic glycolysis* or *the Warburg effect.*

Warburg proposed that cancer cells have the ability to switch to a total reliance on glycolysis for energy when deprived of oxygen. This happens during the initial stages of tumor development, when blood vessels, growing at a much slower rate than the tumor, are unable to supply oxygen to the cancerous cells. In a sense, the cancer cells are dying for oxygen. This condition is called *hypoxia,* and cancer cells under these conditions turn to lactic acid fermentation to maintain ATP production. They also release a compound called *hypoxia inducible transcription factor-I* which increases the production of the enzymes used in glycolysis. Remarkably, once oxygen is restored to the tumor, the cancer cells continue to rely exclusively on glycolysis for ATP production even in the presence of oxygen.

You will recall that yeast cells, under anaerobic conditions, have an increased uptake in sugar. It turns out that cancer cells do this as well. Tumors will have higher glucose uptake relative to normal tissue. The higher the glucose uptake by the tumor, the worse the situation. A higher glucose uptake indicates that the tumor is much more aggressive, and the outlook for the patient is not good.

The increased sugar uptake by cancer cells allows tumors to be imaged by a process called *positron emission tomography* (PET). In this procedure, a radioisotope is injected into the body that undergoes *beta* (β) *decay.* This process is accompanied by the emission of a *positron,* which is the antiparticle of the electron. It has the same mass as the electron but opposite charge. When an antiparticle encounters its corresponding particle, they will annihilate producing gamma (γ) *radiation.* A positron does not have to travel very far in the body until it encounters an electron, and 2 γ-ray photons are produced when the positron and electron annihilate. These γ-ray photons can then be picked up by a detector, and used to generate an image of the body.

To monitor glucose uptake, radioisotopes of carbon, nitrogen, and oxygen are incorporated into glucose. The glucose functions as a radioactive tracer because

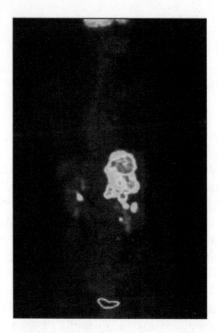

Figure 10-20   An example of how tumors show up in a PET scan of the human body. The brain also shows up due to its exclusive use of glucose and high rate of glucose absorption. (*Courtesy of Wikipedia.*)

through the process of γ-ray emission, where glucose ends up in the body is easily detected. Since tumors take up glucose at a much higher rate than normal tissue, their location and extent can be imaged by a PET scanner. An illustration of this is shown in Fig. 10-20.

The knowledge that cancer cells rely exclusively on glycolysis for energy production may lead to novel cancer treatments. Recently, three cancer researchers named Valeria Flantin, Julie St. Pierre, and Philip Leder showed that a possible cancer treatment could be based on blocking the enzyme lactose dehydrogenase[1]. In their study, which was based on mice with cancer, two groups of mice were used. A control group had cancer but received no treatment. The other group of mice received the treatment blocking lactose dehydrogenase, which shut off the ability of the cancer cells to regenerate NAD^+, shutting off the glycolysis cycle. While all of the mice in the control group died by 4 months, 80% of the mice receiving the treatment were still alive after the 4-month period.

---

[1]Valeria R. Flantin, Julie St-Pierre, and Phillip Leder, "Attenuation of LDH-A expression uncovers a link between glycolysis, mitochondrial physiology, and tumor maintenance." *Cancer Cell*, Vol. 9, No. 6, 425–34 (2006).

Summary

Glycolysis is a process used by cells to extract energy from glucose in the absence of oxygen. It requires the investment of 2 ATP molecules, 10 enzymes, and 2 NAD^+ molecules. When completed, a glucose molecule is converted into 2 molecules of pyruvate with a net gain of 2 ATP molecules.

To continue utilizing glycolysis, the supply of NAD^+ must be replenished. One way this can be done is by lactate fermentation, a process in which NADH is reduced to NAD^+ with the conversion of pyruvate into lactate. In the Cori cycle, the skeletal muscles and liver join together allowing muscle cells to continue to rely on glycolysis for energy. Lactate is released into the blood stream by the muscle cells, where it is taken up by the liver and converted back into glucose. The glucose is then released back to the blood stream where it can be used by the skeletal muscles in glycolysis to extract more energy. In yeast, ethanol fermentation is used to replenish the supply of NAD^+.

Other sugars can be incorporated into glycolysis by conversion into one of the intermediate compounds used in the cycle. Some sugars that can be incorporated in this way include galactose, fructose, and mannose.

Finally, cancer cells rely exclusively on glycolysis for energy. It is believed this occurs because in the early stages of tumor development, cancer cells are starved of oxygen. New cancer treatments may be possible by blocking the enzyme lactate dehydrogenase, depriving the cancer cells of a supply of NAD^+, stopping glycolysis, and killing the cells.

Quiz

1. Hexokinase
   (a) is primarily used in the liver to initiate glycolysis.
   (b) is only used by neurons to initiate glycolysis.
   (c) is used to trap glucose for glycolysis.
   (d) None of the above.

2. Glucokinase
   (a) is used by skeletal muscle to initiate glycolysis.
   (b) is used by the liver, and has a low affinity for glucose.
   (c) is used by the liver, and has a high affinity for glucose.
   (d) is not an isozyme of hexokinase.

3. The NET production of ATP in glycolysis is

(a) 2 ATP molecules.

(b) 4 ATP molecules.

(c) 6 ATP molecules.

(d) 38 ATP molecules.

(e) None of the above.

4. The initial investment of ATP required in glycolysis is

(a) 4 ATP molecules.

(b) 2 ATP molecules.

(c) 1 ATP molecule.

(d) No ATP is required.

5. If a cell needs to continue using glycolysis for energy, it must replenish its supply of

(a) NAD^+ molecules.

(b) NADH molecules.

(c) Protons.

(d) H^+ molecules.

6. Lactate fermentation

(a) is used by yeast to replenish NAD^+.

(b) is used by animal cells to replenish NAD^+.

(c) is used by animal cells to replenish NADH.

(d) is used by animal cells to replenish NAD^+, under aerobic conditions.

7. In the pay-off phase of glycolysis, the total number of ATP molecules produced is

(a) 1.

(b) 2.

(c) 4.

(d) 6.

8. Low levels of ATP in a cell

(a) activate phosphofructo-kinase by driving it to the L conformation.

(b) are correlated with high levels of ADP and AMP, which activate phosphofructo-kinase.

(c) only activate phosphofructo-kinase if levels of AMP are constant.

(d) have no effect on phosphofructo-kinase.

9. Fructose can be utilized in glycolysis

    (a) by entry at step 3 in muscle cells, and step 5 in neurons.

    (b) by entry at step 5 in muscle cells, and step 3 in liver cells.

    (c) by entry at step 2 in muscle cells, and step 5 in liver cells.

    (d) by entry at step 3 in muscle cells, and step 5 in liver cells.

10. The Warburg effect describes

    (a) the tendency of cancer cells to rely on glycolysis, even under aerobic conditions.

    (b) the tendency of cancer cells to quickly abandon glycolysis once oxygen is reintroduced.

    (c) the heavy reliance of cancer cells on lactate dehydrogenase.

    (d) None of the above.

CHAPTER 11

The Citric Acid Cycle and Oxidative Phosphorylation

An important if not crucial step in the evolution of complex organisms was the incorporation of mitochondria into eukaryotic cells. This step increased the energy available to cells by a dramatic amount, because mitochondria are energy powerhouses. Mitochondria, which are believed to have once been free-living organisms and have their own genetic component, are essential for liberating the maximum possible energy from carbohydrates, fatty acids, and amino acids. This is done in a two step process. The first step is the oxidation of the acetyl group of acetyl-CoA, a product of catabolism of carbohydrates, fatty acids, and many amino acids. This is done in a process called the *citric acid cycle* [or tricarboxylic acid (TCA) cycle], which generates 1 guanine triphosphate (GTP), 3 nicotinamide adenine dinucleotide (NADH),

and 1 flavin adenine dinucleotide ($FADH_2$) molecule. In addition, 2 CO_2 molecules are produced.

The products of the citric acid cycle can be utilized to produce ATP molecules. This process, which we also consider in this chapter, is called *oxidative phosphorylation*. In oxidative phosphorylation reduced nicotinamide and flavin nucleotides produced by the oxidation of carbohydrates and lipids are oxidized to synthesize ATP.

The Citric Acid Cycle

We begin our discussion of the citric acid cycle with the production of *acetyl-CoA*, which is the fuel used to drive the cycle. For example, acetyl-CoA is produced from pyruvate using the enzyme pyruvate dehydrogenase, coenzyme A (CoASH), and one NAD^+. The product of this reaction is acetyl-CoA with the liberation of 1 CO_2 molecule + NADH.

The acetyl-CoA molecule is the input molecule for the citric acid cycle, but *oxaloacetate* is also required. The latter molecule is not considered "fuel" for the cycle because it is a recycled product used in the reaction. We can summarize the essential facts of the citric acid cycle in the following way:

- The citric acid cycle is an eight-step reaction.
- It requires 8 enzymes.
- The final product is oxaloacetate.
- Two NADH molecules are produced.
- One GTP molecule is produced.
- One $FADH_2$ molecule is produced.
- The cycle is accompanied by the liberation of 2 CO_2 molecules.

The eight enzymes utilized in the citric acid cycle are:

1. Citrate synthase
2. Aconitase
3. Isocitrate dehydrogenase
4. α-Ketoglutarate dehydrogenase
5. Succinyl-CoA synthetase
6. Succinate dehydrogenase
7. Fumarase
8. Malate dehydrogenase

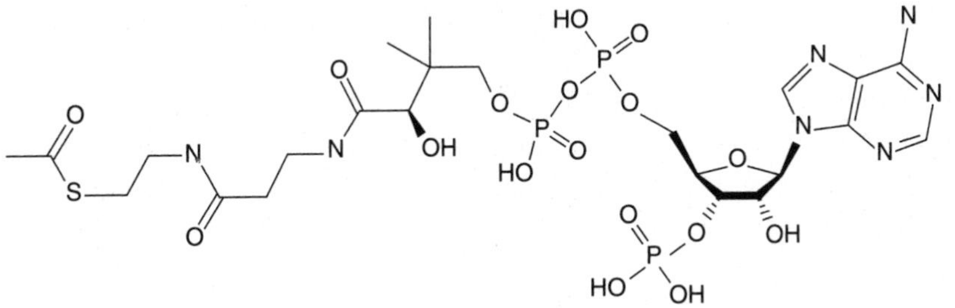

Figure 11-1   Acetyl-CoA, the input or "fuel" for the citric acid cycle.

The production of acetyl-CoA for the citric acid cycle must occur inside the mitochondria because it is unable to cross the mitochondrial membrane from the cytosol. As a result, other molecules must be utilized. It turns out that pyruvate, fatty acids, and some amino acids are able to cross the mitochondrial membrane where they can be used by the mitochondria to produce acetyl-CoA.

STEPS IN THE CITRIC ACID CYCLE

Now let's consider the eight steps of the citric acid cycle.

Step 1: Acetyl-CoA → Citrate

In the first step of the cycle, the enzyme citrate synthase catalyzes the condensation of acetyl-CoA (Fig. 11-1) with oxaloacetate (Fig. 11-2). This produces a molecule called *citrate,* and liberates coenzyme A. One water molecule is also required. Note that citrate is a tertiary alcohol that cannot be oxidized very easily (Fig. 11-3), so it must be further processed. This is done in the second step of the cycle.

Step 2: Citrate → Isocitrate

In step 2 of the cycle, there are two substeps used to generate isocitrate, which is easier to oxidize. First, the enzyme aconitase catalyzes the dehydration of the citrate

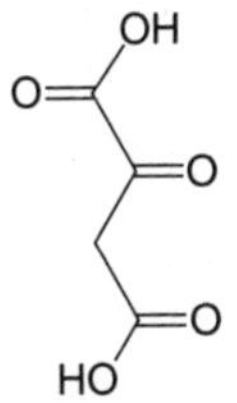

Figure 11-2   Oxaloacetate, a molecule required in the first step of the citric acid cycle and produced in the final or eighth step.

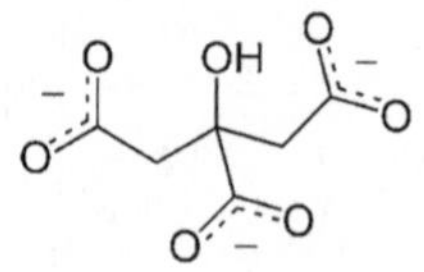

Figure 11-3 The citrate molecule.

molecule producing an intermediate molecule called *cis-aconitate*. The cis-aconitate molecule, remaining bound to the enzyme, is then hydrated to produce the product of the second step, isocitrate. The net effect of this reaction is that the hydroxyl group on citrate is moved from the third carbon atom to the fourth carbon atom, producing an isomer of citrate that is easier to oxidize.

Step 3: Isocitrate → α-Ketoglutamate

The next step in the citric acid cycle also has two substeps or phases. First, isocitrate is oxidized by the enzyme isocitrate dehydrogenase producing *oxalosuccinate*. Like cis-aconitate in step 2, this molecule is an intermediate that never dissociates from the enzyme. Instead it undergoes further processing. In the second phase of step two, the enzyme decarboxylates oxalosuccinate to produce α-ketoglutamate. In step 3, one NADH is produced in the first phase and a molecule of CO_2 is release in the second phase.

Step 4: α-Ketoglutamate → Succinyl-CoA

The next step in the reaction, which is catalyzed by α-ketoglutamate dehydrogenase, decarboxylates α-ketoglutamate producing succinyl-CoA (Fig. 11-4). One NAD^+ and 1 molecule of coenzyme A are required for this step of the reaction, which produces 1 NADH and 1 molecule of CO_2.

Step 5: Succinyl-CoA → Succinate

In this step, a high energy GTP molecule is produced. This is substrate level phosphorylation. The reaction is catalyzed by the molecule succinyl-CoA synthetase, and the coenzyme A molecule consumed in the production of succinyl-CoA is released (see Fig. 11-5).

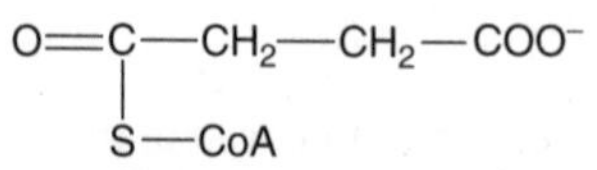

Figure 11-4 Succinyl-CoA.

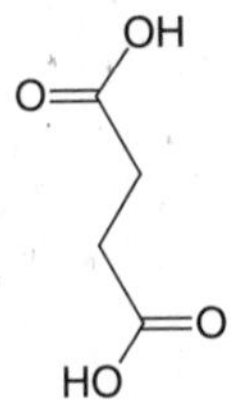

Figure 11-5   Succinate.

Steps 1 to 5 can be considered to be the first part of the citric acid cycle. Two NADH, 1 GTP and 2 CO_2 molecules have been produced. In the second part of the cycle, succinate will be oxidized back to the starting product in the reaction, oxaloacetate, which can then be used in another round of the cycle. This oxidation requires three steps and produces 1 $FADH_2$ and 1 NADH molecule.

Step 6: Succinate → Fumarate

In this step, succcinate dehydrogenase oxidizes the succinate molecule producing fumarate. One FAD is utilized in this reaction, and succinate dehydrogenase eliminates 2 hydrogens in the oxidation of the central single bond of the succinate molecule. The coenzyme FAD is therefore reduced to $FADH_2$ in the process (see Fig. 11-6).

Step 7: Fumarate → L-Malate

In this step, fumarate hydratase or *fumarase* catalyzes a reversible reaction in which fumarate is transformed into malate. One molecule of water is required, in a process which transforms the central double bond of fumarate into a single bond. The top carbon of the double bond is changed from CH → CHOH and the bottom carbon of the double bond is transformed from CH → CH_2 (see Fig. 11-7).

Step 8: Malate → Oxaloacetate

We have now reached the final step in the cycle. In this last step, the molecule oxaloacetate is regenerated from malate, allowing the cycle to start anew (provided there is a supply of acetyl-CoA). The enzyme which catalyzes this step is malate dehydrogenase, and 1 NADH molecule is produced.

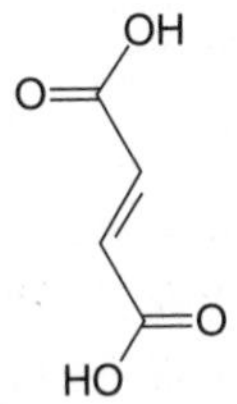

Figure 11-6   Fumarate.

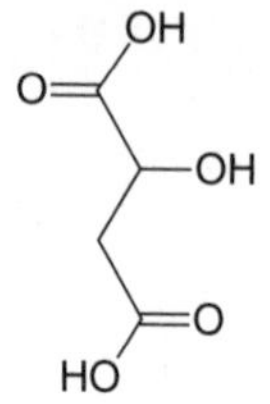

Figure 11-7 Malate.

NOTE: The only input molecule to the citric acid cycle is the acetyl-CoA molecule, whose carbon atoms are lost as CO_2 molecules. All other molecules in the cycle are regenerated intermediates. Acetyl-CoA is the *fuel* for the cycle which must be input from the outside.

ENERGETICS OF THE CITRIC ACID CYCLE

The citric acid cycle is an *exergonic* reaction with

$$\Delta G^{\circ\prime} = -60 \text{ kJ/mol}$$

Two steps in the reaction are *endergonic*. These are:

- Step 2, conversion of citrate to isocitrate with $\Delta G^{\circ\prime} = +5$ kJ/mol.
- Step 8, the conversion of malate to oxaloacetate with $\Delta G^{\circ\prime} = +30$ kJ/mol.

The most exergonic steps in the cycle are

- Step 1, the conversion of acetyl-CoA to citrate with $\Delta G^{\circ\prime} = -32$ kJ/mol.
- Step 4, the conversion of α-ketoglutamate to succinyl-CoA, with $\Delta G^{\circ\prime} = -30$ kJ/mol.

Since they are endergonic, steps 2 and 8 of the citric acid cycle are *unfavored*. It is possible for them to proceed because they are followed by the energetically favorable steps 3 and 1. In step 2, at equilibrium, the reversible reaction which converts citrate to isocitrate actually favors the formation of citrate. However, step 3 of the reaction is exergonic and hence is energetically favorable. So when isocitrate is formed, it is removed and utilized in step 3.

Similarly, at equilibrium the reversible reaction which converts malate to oxaloacetate favors the formation of malate. However oxaloacetate is an input molecule for step 1 of the cycle, which is exergonic (and hence energetically favorable). So when oxaloacetate forms then step 1 of the cycle will proceed removing the molecule.

CONTROL OF THE CITRIC ACID CYCLE

The citric acid cycle is controlled by two factors:

- Regulatory enzymes
- The energy requirements of mitochondria

There are four regulatory enzymes in the citric acid cycle. These are

1. Citrate synthase
2. Isocitrate dehydrogenase
3. 2-Oxogulta dehydrogenase
4. Succinate dehydrogenase

Citrate synthase catalyzes the first step of the citric acid cycle, which is the condensation reaction of acetyl-CoA and oxaloacetate using an induced fit process. Citrate synthase has an oxaloacetate binding site. Initially, the enzyme exists in an *open* state with a cleft where we find the binding site. It first binds a molecule of oxaloacetate to this site. Upon binding, oxaloacetate induces a change in the structure of the citrate synthase enzyme into a *closed* state that produces a binding site for acetyl-CoA. This is the induced fit aspect of the action of this enzyme.

An intermediate compound is formed from acetyl-CoA and oxaloacetate called *citroyl-CoA*. This molecule remains bound to the enzyme, and the hydrolysis of a thioester bond results in the production of citrate and coenzyme A. The first step in the citric acid cycle is inhibited by NADH and succinyl-CoA. In the first case, energy requirements of mitochondria are a limiting factor of the initiation of the citric acid cycle. Succinyl-CoA can also bind to the active acetyl-CoA binding site and so competes with acetyl-CoA in the initiation of the first step.

Step 2 of the citric acid cycle is an important regulatory step as well. This is because the enzyme which catalyzes the reaction, isocitrate dehydrogenase, is an allosteric enzyme whose activity is inhibited by the presence of high-energy compounds. That is, it is inhibited by the presence of ATP and NADH, and conversely it is activated by ADP and NAD^+. When compounds with high-energy bonds are present, isocitrate dehydrogenase is inhibited, leading to a "shut down" process that works as follows. Without the action of the enzyme, isocitrate begins to accumulate in the mitochondria. This leads to an equilibrium situation, which favors the conversion of isocitrate to citrate, with 7% isocitrate and 93% citrate at equilibrium. Citrate begins to accumulate in the mitochondria but a transport protein in the membrane of the mitochondria allows it to exit to the cytoplasm. Here it acts to inhibit the enzymes pyruvate kinase and phosphofructokinase, essentially shutting off metabolism.

Isocitrate dehydrogenase is dependent on NAD^+ for its activity. In addition, it requires a cofactor which can be either Mn^{2+} or Mg^{2+}.

The next regulatory enzyme in the citric acid cycle is 2-oxoglutarate dehydrogenase, which is inhibited by high levels of NADH and succinyl-CoA. Hence a high energy state in the mitochondria inhibits the action of this enzyme, which acts in step 4 of the citric acid cycle to produce succinyl-CoA.

The final regulatory enzyme in the citric acid cycle is succinate dehydrogenase. This enzyme acts on step 6, where succinate is converted into fumarate with the transformation (reduction) of an FAD molecule to $FADH_2$. The enzyme is actually inhibited by the last step in the cycle, because an accumulation of oxaloacetate inhibits the enzyme. As such we say that a *feedback loop* exists between steps 8 and 6 of the cycle, controlling the production of fumarate when oxaloacetate levels get too high. As the energy levels of the mitochondria go up with the correlated increase in oxaloacetate concentration, succinate dehydrogenase is inhibited putting the breaks on the citric acid cycle. An interesting feature of succinate dehydrogenase is that an FAD molecule is permanently attached to the enzyme via a covalent bond to an HIS residue of the enzyme. The FAD molecule acts as the electron acceptor in step 6 of the citric acid cycle. Oxidation of an alkane to an alkene is accomplished by FAD, and succinate is an alkane, while fumarate is an alkene. Coenzyme Q reoxidizes succinate dehydrogenase in the electron transport chain.

THE CITRIC ACID CYCLE AND ATP GENERATION

In the end, the products of the citric acid cycle lead to the production of ATP. In short, 12 ATP molecules are produced. This is accomplished by the use of 8 electrons (four pairs) from the acetyl-CoA molecule that enters the citric acid cycle. We can summarize the citric acid cycle by saying that acetyl-CoA is oxidized to 2 CO_2 molecules and, in the process, 3 molecules of NAD^+ are reduced to 3 molecules of NADH. This process uses up 6 out of the 8 electrons (or 3 of the 4 electron pairs) available at the start of the cycle. Recalling that in step 6 one FAD is reduced to $FADH_2$, we account for the remaining pair of electrons. These electron pairs are utilized in oxidative phosphorylation to produce ATP molecules. The 3 electron pairs in the NADH molecules are used to generate 3 ATP molecules and the $FADH_2$ molecule is used to generate 2 ATP molecules.

Cataplerotic Reactions

Cataplerotic reactions or *anabolic reactions* are processes which utilize the intermediate products in the citric acid cycle. While at first glance this appears to be a bad thing for the citric acid cycle and the energy requirements of the mitochondria, it is actually beneficial because it prevents the excessive accumulation of intermediate products in the mitochondria. There are four anabolic reactions that we consider here.

1. Glucose Production

Malate (the product of step 7) can be transported across the mitochondrial membrane to the cytosol. There, it is converted into oxaloacetate and used in gluconeogenesis by the cell. The oxaloacetate produced in the citric acid cycle is not utilized in this fashion because it cannot cross the mitochondrial membrane.

2. Synthesis of Amino Acids

In the presence of glutamate dehydrogenase, and NADH or NADPH, α-ketoglutamate can be converted into the amino acid glutamate with the liberation of a water molecule and $NAD(P)^+$. In addition, oxaloacetate can be transformed into aspartate.

3. Production of Fatty Acids

Citrate can cross the mitochondrial membrane into the cytosol, where it is utilized to generate acetyl-CoA. This acetyl-CoA is then used to biosynthesize fatty acids and cholesterol.

4. Porphyrin Metabolism

Heme precursors called *porphyrins* are important components in the synthesis of heme proteins, hemoglobin, and cytochromes. The first step in the synthesis of a porphyrin molecule requires glycine and succinyl-CoA, plus a Mg^{2+} cofactor. This results in the production of a molecule called α-aminolerulinate, CO_2, and CoASH. The succinyl-CoA used in this reaction is made available by step 4 of the citric acid cycle. The reactions in the citric acid cycle are summarized in Fig. 11-8.

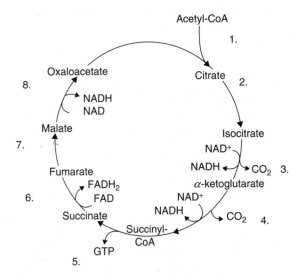

Figure 11-8 The citric acid cycle.

Oxidative Phosphorylation

Ultimately, we are interested in the production of high-energy ATP molecules, utilizing oxygen in the mitochondria. ATP synthesis is accomplished in a process called *oxidative phosphorylation*, which relies on *electron transport* to obtain the required energy and proton gradient. The *electron transport chain* uses the electron pairs carried by the NADH and $FADH_2$ molecules that were reduced in the citric acid cycle.

In the citric acid cycle, which can be considered an intermediate step in metabolism, electrons are transferred to the reduced electron carriers NADH and $FADH_2$. In the electron transport chain, these electrons are transferred to oxygen and protons to give H_2O and make available the energy required for ATP synthesis. This is possible because the oxidation of NADH by O_2 can be coupled to ATP synthesis and it is energetic enough to generate large amounts of ATP. To see why, we can examine the standard reduction potential E_0' for two reactions involving electron transfer. For $NAD^+/NADH$ the electron transport reaction is:

$$NAD^+ + H^+ + 2e^- \rightarrow NADH$$

The standard reduction potential of this reaction is:

$$E_{0(donor)}' = -0.32 \text{ V}$$

(we are labeling this as *donor* because later we will consider this as a half-reaction which donates two electrons for the next half-reaction). Now, consider the transfer of two electrons in the reaction

$$\frac{1}{2}O_2 + 2H^+ + 2e^- \rightarrow H_2O$$

In this case:

$$E_{0(acceptor)}' = +0.82 \text{ V}$$

The larger potential for this reaction indicates that when coupled to electron transport by $NAD^+/NADH$, the O_2/H_2O reaction will have a *larger affinity* for the electrons. That is, NADH will donate 2 electrons in the following reaction:

$$\frac{1}{2}O_2 + H^+ + NADH \rightarrow H_2O + NAD^+ \tag{11-1}$$

The *overall* standard reduction potential for this reaction is calculated by taking the difference between $E_{0(\text{acceptor})}'$ and $E_{0(\text{donor})}'$:

$$\Delta E_0' = E_{0(\text{acceptor})}' - E_{0(\text{donor})}' = 0.82 - (-0.32) = 1.14 \text{ V}$$

This value can be used to calculate the standard free energy for the reaction described in Eq. 11-1, using the formula

$$\Delta G_0' = -nF\Delta E_0'$$

Here, n is the number or electrons transferred per mol, and F is the *Faraday constant* which is given by $F = 96{,}485$ C/mol e^- (where C are coulombs). Noting that the standard reduction potential is given in volts which are defined as $1 \text{ V} = 1 \text{ J/C}$, in this case the standard free energy is then:

$$\Delta G_0' = -2\frac{\text{mol } e^-}{\text{mol}_{\text{reactant}}} \times 96{,}485 \ \frac{\text{C}}{\text{mol } e^-} \times 1.14 \text{ J/C}$$

$$= -218 \text{ kJ/mol}_{\text{reactant}}$$

Since this value is *negative,* this means that 218 kJ of energy is released per mol of reactants in Eq. 11-1. To synthesize ATP via $\text{ADP} + \text{P}_i \rightarrow \text{ATP}$, about 30.5 kJ/mol is required. Hence more than enough energy is available to drive the reaction which synthesizes ATP.

Other Components of the Electron Transport Chain

In the citric acid cycle, we also noted the transfer of electrons to the flavin nucleotide in the reaction:

$$\text{FAD} + 2\text{H}^+ + 2e^- \rightarrow \text{FADH}_2$$

Another reaction involving the flavin mononucleotide FMN is also possible

$$\text{FMN} + 2\text{H}^+ + 2e^- \rightarrow \text{FMNH}_2$$

Earlier, we have seen that FADH_2 is generated in step 6 of the citric acid cycle, where:

$$\text{succinate} + \text{FAD} \rightarrow \text{fumarate} + \text{FADH}_2$$

In that discussion, we noted that the FAD molecule is covalently bound to the enzyme succinate dehydrogenase. This type of arrangement is true in general for the flavin nucleotides; they are not free in solution but are covalently bound to a dehydrogenase. These enzymes are in turn bound to the inner membrane of the mitochondria, where they can participate in the electron transport chain.

$FADH_2$ can also be synthesized from the transfer of electrons from glycerol 3-phosphate in fatty acid oxidation

$$Glycerol\ 3\text{-phosphate} + FAD \rightarrow Dihydroxyacetone\ phosphate + FADH_2$$

We will see in a moment how the $FADH_2$ molecule is utilized in the electron transport chain. Now let's turn to the next component utilized in this process, *coenzyme Q*, a molecule sometimes called *ubiquinone* because it is "ubiquitous" in organisms that utilize oxygen in metabolism. It participates in an electron transfer reaction to form $CoQH_2$, which is known as *ubiquinol* as follows:

$$CoQ + 2H^+ + 2e^- \rightarrow CoQH_2$$

The final component used in the electron transport chain that we will note is *cytochrome C*. This molecule is a *heme protein* that is found on the inner membrane of the mitochondria. Cytochromes are proteins that have a prosthetic heme group. There are three classifications of cytochromes found in mitochondria, denoted as a, b, and c. The designation is made based on the type of heme group contained in the protein. We will see in a moment that it plays a role in the electron transport chain because it is able to undergo oxidation and reduction reactions.

The Electron Transport Chain and Protein Complexes

The function of the electron transport chain is to liberate the free energy that can be used to synthesize ATP and to create a proton gradient. This is done by oxidizing NADH and $FADH_2$ in a process whereby four *protein complexes* denoted as *complex I, complex II, complex III, and complex IV* are used to catalyze oxidation reactions and pass electrons from lower to higher standard reduction potentials. Summarizing what we have discussed so far, the electron transport chain utilizes the following molecules:

- NADH
- $FADH_2$
- Coenzyme Q
- Cytochrome c
- Oxygen

Table 11-1   The Protein Complexes of the Electron Transport Chain

Complex	Description or Name	$\Delta E_0'$	$\Delta G^{0\prime}$
I	NADH/CoQ oxidoreductase	0.36 V	−69.5 kJ/mol
II	Succinate/CoQ reductase	—	—
III	CoQ cytochrome c oxidoreductase	0.19 V	−36.7 kJ/mol
IV	Cytochrome c oxidase	0.58 V	−112 kJ/mol

The protein complexes I to IV catalyze the reactions which take place. A summary of the protein complexes along with the standard reduction potential and change in free energy for the reactions they catalyze is shown in Table 11-1.

In the electron transport chain, coenzyme Q serves to transport electrons from complex I to complex II, while cytochrome c serves to transport electrons from complex III to complex IV. We now consider the electron transport chain, which can be viewed as operating in four distinct processes, each mediated by a different protein complex.

COMPLEX I: OXIDATION OF NADH

The first step in the electron transport chain is catalyzed by complex I. In this step, electrons are transferred from NADH to coenzyme Q. In addition, complex I moves four protons across the mitochondrial membrane, setting up the proton gradient. The reaction catalyzed is:

$$NADH + CoQ \rightarrow NAD^+ + CoH_2$$

Two facts should be noted about this reaction. The first is that ubiquinol CoH_2 is a *lipid-soluble electron carrier* and can diffuse within the mitochondrial membrane. Secondly, the change in standard free energy for this reaction is $\Delta G^{0\prime} = -69.5$ kJ/mol, which is well above the 30.5 kJ/mol required for the synthesis of ATP.

COMPLEX II: OXIDATION OF FADH$_2$

The next leg in the electron transport chain is mediated by protein complex II. In this case, the protein mediates the transfer of electrons from succinate dehydrogenase to coenzyme Q. Recalling that FAD is covalently bound to succinate dehydrogenase, the reaction that actually takes place is:

$$FADH_2 + CoQ \rightarrow FAD + CoQH_2$$

The change in the standard free energy in this case is $\Delta G^{0'} = -16.4$ kJ/mol, which is not sufficient energy to synthesize ATP. Moreover, complex II does not move any protons across the membrane and does not contribute to the proton gradient. The purpose of this step is only to transfer electrons into the chain.

COMPLEX III: OXIDATION OF COQ

In this step, coenzyme Q is oxidized. This is done in a reaction catalyzed by protein complex III, which transfers electrons from ubiquinol to cytochrome c:

$$CoQH_2 + 2 \text{ cytochrome c} \rightarrow CoQ + 2 \text{ cytochrome c (reduced)}$$

The change in the standard free energy in this reaction is $\Delta G^{0'} = -36.7$ kJ/mol. While this is not as large as what we saw in the reaction catalyzed by protein complex I, it is sufficient for the synthesis of ATP (see Fig. 11-9). Protein complex III

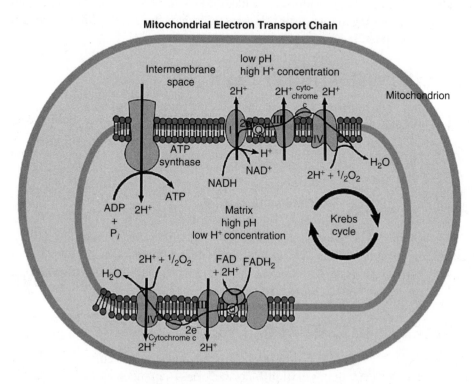

Figure 11-9 A schematic representation of the Krebs (citric acid) cycle, the electron transport chain, and ATP synthesis in mitochondria. (*Courtesy of Wikipedia.*) Created by Rozzychan, http://en.wikipedia.org/wiki/Image:Etc2.png. Licensed under the share alike creative commons license, http://creativecommons.org/licenses/by-sa/2.5/.

moves 2 protons across the membrane. Cytochrome c is a water-soluble protein that is found in the intermembrane space between the inner and outer mitochondrial membranes. It is loosely bound to the outer surface of the inner mitochondrial membrane, and it uses the heme for electron transport. Reduced hemes are highly reactive and efficient at electron transport. In fact, by themselves they are too efficient and reactive. The function of the protein component of cytochrome c is to dampen this activity. Electron transport is actually most efficiently accomplished in these molecules moving the electrons through bonds using a process called *quantum mechanical tunneling.* This is a process whereby a particle such as an electron can move past a potential barrier that would forbid movement in classical mechanics. This is possible because of the wavelike behavior of the electron, which allows it to leak past barriers and appear at points on the other side of the barrier. Van der Waals interactions arbitrate electron tunneling across protein–protein barriers.

COMPLEX IV: OXIDATION OF CYTOCHROME C

Complex IV catalyzes the transfer of electrons from cytochrome c to O_2. The end product of the reaction is water. It can be written as:

$$2 \text{ cytochrome c (reduced)} + \frac{1}{2}O_2 \rightarrow 2 \text{ cytochrome c (oxidized)} + H_2O$$

The change in standard free energy in this case is the largest one in the chain: $\Delta G^{0\prime} = -112 \text{ kJ/mol}$, which is more than enough to synthesize ATP. Complex IV transfers 2 protons across the mitochondrial membrane, contributing to the proton gradient.

Interestingly, while the electron transport chain is vital for the high energy requirements of complex organisms, it may be involved in the initiation of disease processes and the affects of aging.

ATP Synthesis

In addition to gaining or shall we say obtaining the free energy needed to synthesize ATP, the electron transport chain serves to set up a proton (H^+) gradient. This gradient exists between the intermembrane space and the inside of the mitochondria (the *mitochondrial matrix*). We can characterize this gradient in the following way:

- A high proton (H^+) concentration exists in the intermembrane space.
- A low proton (H^+) concentration exists in the mitochondrial matrix.

- Protons are pumped from inside across the inner mitochondrial membrane by the protein complexes I, III, and IV in the electron transport chain.

- This establishes an *electrochemical gradient* across the inner mitochondrial membrane.

There is but one purpose for establishing this proton gradient. It allows the free energy liberated in the electron transport chain to be stored in an electrochemical gradient called the *proton motive force* (PMF). This is analogous to the storage of electrical energy in a battery, which is done using an *electromotive force*. Note that:

- A region with a high H^+ concentration is a region with a high electrical potential.

- A region with a low H^+ concentration is a region with a low electrical potential.

The return of protons to the mitochondrial matrix, which is accomplished by *proton translocating ATP synthase* or *complex V,* provides the energy that directly drives ATP synthesis.

Some facts to keep in mind about oxidative phosphorylation:

- Oxidation of NADH results in the production of 3 ATP molecules.

- Oxidation of succinate results in the production of 2 ATP molecules.

- Two to three protons are passed *back* to the mitochondrial matrix by complex V (ATP synthase) for each ATP synthesized.

ATP synthase has globular units called F_1 complexes that protrude from the matrix side of the inner mitochondrial membrane. The F_1 complexes are also called ATPase, and they have subunits that can hydrolyze ATP but not synthesize it. The subunits that make up the F_1 complex are five polypeptide chains denoted α, β, γ, δ, and ε. The δ- and ε-chains serve to bind the F_1 complex to the mitochondrial membrane. The α- and β-chains serve as the site of enzyme activity.

The F_1 complex is bound to membrane channels called the F_0 *complex*. The F_0 complex functions as a proton pump which brings protons back from the intermembrane space to the mitochondrial matrix. The F_0 works in concert with the F_1 subunits to produce ATP. Three β-subunits appear to work together, one binds ADP, one binds ATP, while a third is empty. This is not a static situation, rather, each subunit cycles through different states with the transformation between each state made possible via the energy obtained from proton transport. Hence

- Conformational changes in ATPase driven by proton transport appear to drive the synthesis of ATP.

Protons are transported through the F_0 complex causing conformational changes in the three β-subunits. Three states of a β-subunit appear possible. These are denoted the *tight* (T) state, the *loose* (L) state, and the *open* (O) state. These states work as follows:

- Starting in the L state, ADP + P$_i$ are bound to the β-chain.

- Proton translocation across the membrane provides energy, which causes a conformational change in the β-chain, changing the L state to the T state.

- In the L state, ADP and inorganic phosphate can bind forming ATP.

- A neighboring β-chain will be in the T state containing an ATP molecule. The conformational change described above in which the L state changes to the T state induces this β-chain to transform to the O state. Then the synthesized ATP molecule is released.

Once ATP has been synthesized, it is transported out of the mitochondria into the cytoplasm. In addition, ADP is transported into the mitochondria so that more ATP can be synthesized as needed.

Summary

The citric acid cycle provides the free energy necessary for ATP synthesis by reducing NAD$^+$ and FAD molecules. Much to the dismay of global warming guardians, this process liberates CO$_2$. More specifically, electrons are transferred from acetyl-CoA in the citric acid cycle to produce NADH and FADH$_2$, which then feed the electron transport chain. The electron transport chain serves two purposes—to provide the free energy necessary for ATP synthesis and to create a proton gradient that stores that energy as a PMF. The energy is then utilized by ATP synthase to produce ATP from ADP and inorganic phosphate in the mitochondrial matrix. The ATP is then exported to the cell cytoplasm where it can be used to drive energy intensive processes.

Quiz

1. In the first step of the citric acid cycle
   - (a) acetyl-CoA is bound to the enzyme citrate synthase, and then it undergoes a condensation reaction with oxaloacetate.
   - (b) acetyl-CoA undergoes a free condensation reaction with oxaloacetate.
   - (c) oxaloacetate is bound to the enzyme citrate synthase and then it undergoes a condensation reaction with acetyl-CoA.
   - (d) oxaloacetate is bound to the enzyme acetate dehydrogenase, and then it undergoes a condensation reaction with acetyl-CoA.

2. In the citric acid cycle

    (a) all steps are exergonic.

    (b) all steps except 2 are exergonic.

    (c) all steps except 2 are endergonic.

    (d) all steps are endergonic.

3. The final intermediate product in the citric acid cycle is

    (a) L-malate.

    (b) Acetyl-CoA.

    (c) Oxaloacetate.

    (d) Fumarate.

4. GTP is synthesized in the citric acid cycle

    (a) in the transformation of succinyl-CoA to succinate.

    (b) in the transformation of succinate to fumarate.

    (c) in the transformation of succinate to succinyl-CoA.

    (d) GTP is *not* synthesized in the citric acid cycle, 1 molecule of ATP is.

5. Isocitrate dehydrogenase

    (a) is activated by high concentrations of ATP and NADH.

    (b) is activated by high concentrations of ATP and NADPH.

    (c) is unaffected by high concentrations of NADPH.

    (d) is inhibited by high concentrations of high-energy compounds.

6. The fuel of the citric acid cycle could be described as

    (a) acetyl-CoA plus oxaloacetate, which participate in the first step of the cycle.

    (b) only acetyl-CoA, oxaloacetate is a recycled intermediate.

    (c) acetyl-CoA and H_2O.

    (d) acetyl-CoA and GDP.

7. The electron transport chain

    (a) creates a proton gradient across the inner mitochondrial membrane, with a low proton concentration in the mitochondrial matrix.

    (b) creates a proton gradient across the inner mitochondrial membrane, with a low proton concentration in the intermembrane space.

(c) creates a proton gradient across the outer mitochondrial membrane, with a high proton concentration in the intermembrane space.

    (d) creates a proton gradient across the outer mitochondrial membrane, with a low proton concentration in the intermembrane space.

8. The largest amount of ATP production is obtained from

    (a) the oxidation of succinate dehydrogenase.

    (b) the oxidation of NADH molecules.

    (c) the oxidation of $FADH_2$ molecules.

    (d) through the electron chain.

9. In the electron transport chain, complex II

    (a) does not contribute to the proton gradient, but mediates the transfer of electrons from succinate to cytochrome a.

    (b) does not contribute to the proton gradient, but mediates the transfer of electrons from succinate to cytochrome c.

    (c) does not contribute to the proton gradient, but mediates the transfer of electrons from succinate to coenzyme Q.

    (d) contributes to the proton gradient, and mediates the transfer of electrons from succinate to coenzyme Q.

10. The transport of protons back into the mitochondrial matrix appears to

    (a) cause a conformational change in a β-subunit that changes it from the L to the T state allowing ADP and inorganic phosphate to bind forming ATP.

    (b) cause a conformation change in a γ-subunit that changes it from the L to the T state allowing ADP and inorganic phosphate to bind forming ATP.

    (c) cause a conformational change in a β-subunit that changes it from the T to the O state releasing all ATP molecules.

    (d) cause a conformational change in a β-subunit that changes it from the O to the L state allowing ADP and inorganic phosphate to bind forming ATP.

CHAPTER 12

Control of Chemical Processes—Signal Biomolecules

In a multicellular organism, it is necessary for cells to communicate with each other to regulate various processes. While there are some synapses in the brain which operate using direct electrical contact, the primary means of communication in the body is through the use of chemical messengers. There are two basic ways that chemical messengers act to carry signals in the body: via *hormones* and *neurotransmitters*. A hormone is a chemical messenger that carries a signal from one cell to another, often from one part of the body to a cell in a distant location. Neurotransmitters are substances that are released by neurons to control or alter the behavior of their neighbors.

Cells respond to their environment by altering their biochemical processes. For all chemical messengers, be they hormones or neurotransmitters, the interface between the demands of the environment and the inner workings of the cell is accomplished through molecular signals that typically attach to the cell membrane. When a messenger attaches to the cell membrane through signal transduction it affects the biochemistry of the cell. We begin the chapter with a discussion of hormones.

Hormones

A hormone is a chemical that acts as a messenger transmitting a signal from one cell to another. When it binds to another cell which is the target of the message, the hormone can alter several aspects of cell function, including cell growth, metabolism, or other function. Hormones can turn various bioprocesses on or off, and as such play a fundamental role in the livelihood of a multicellular organism.

Hormones can be classified in a wide variety of ways. The first way that we can characterize a hormone is by looking at the distance over which the hormone acts. There are three primary ways we can classify hormones in this way:

- *Autocrine:* A hormone is classified as an autocrine hormone if it acts on the same cell that released it.

- *Paracrine:* A paracrine hormone is one that acts on cells which are nearby relative to the cell which released it. An example of paracrine hormones includes *growth factors,* which are proteins that stimulate cellular proliferation and differentiation. Specifically, consider the binding of white blood cells to T cells. When the white blood cell binds to a T cell, it releases a protein growth factor called *interleukin-1.* This causes the T cell to proliferate and differentiate.

- *Endocrine:* An endocrine hormone is one that is released into the bloodstream by *endocrine glands.* The receptor cells are distant from the source. This is probably how most people think of hormones. An example of an endocrine hormone is *insulin,* which is released by the pancreas into the bloodstream where it regulates glucose uptake by liver and muscle cells.

Hormones can also be classified by the type of molecule the hormone is. There are three major classifications you should be aware of:

- *Steroids:* Steroid hormones are for the most part derivatives of cholesterol.

- *Amino acid derivatives:* Several hormones (and neurotransmitters) are derived from amino acids.

- *Polypeptides:* Many hormones are chains of amino acids.

There are a large number of hormones that act in the body, and unfortunately we will only be able to discuss a small number of them. We now consider each of the major hormone types in turn.

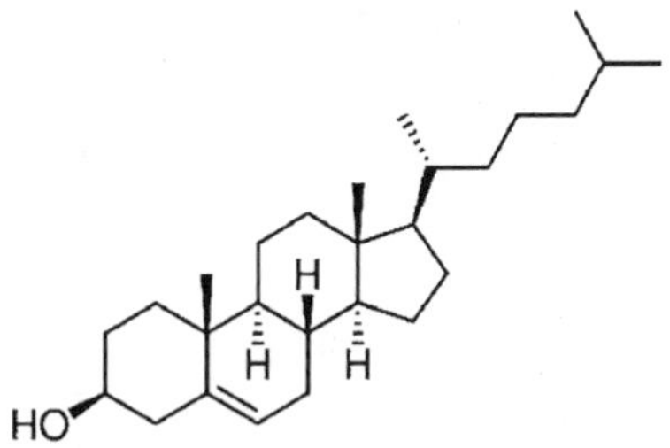

Figure 12-1   Steroid hormones are derived from cholesterol.

STEROID HORMONES

Most steroid hormones are derived from cholesterol (see Fig. 12-1). Cholesterol is converted into a steroid hormone via the cleavage of a 6 carbon residue, producing a compound called *pregnenolone,* which is shown in Fig. 12-2. Pregnenolone is a precursor to C_{18}, C_{19}, and C_{21} steroids. We classify these steroids as follows:

- *Estrane:* An estrane is a steroid consisting of 18 carbon atoms.
- *Androstane:* An androstane is a steroid consisting of 19 carbon atoms.
- *Pregnane:* A pregnane is a steroid consisting of 21 carbon atoms.

EXAMPLES OF STEROID HORMONES

A steroid hormone derived directly from pregnenolone is called *progesterone* (see Fig. 12-3). This hormone, which is secreted primarily by the *corpus luteum,* is a precursor for many other hormones. The corpus luteum is a temporary structure in females, derived from an ovarian follicle during the menstrual cycle. Progesterone is also produced by the adrenal glands. In addition to acting as a precursor molecule in the synthesis of other hormones, progesterone plays several fundamental roles, the most familiar of which deal with the female reproductive system, where it prepares the female body for pregnancy. In ovulating women, it is released during

Figure 12-2   Pregnenolone is a precursor to many hormones.

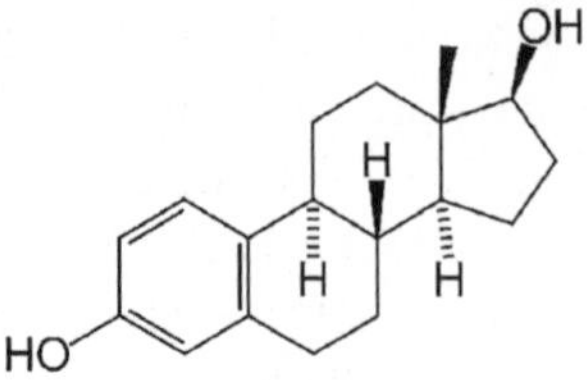

Figure 12-3 Estradiol is a hormone which plays a role in the development of female secondary sexual characteristics, among other functions.

the second 2 weeks of the menstrual cycle. Specifically, three effects of progesterone when it is secreted after ovulation can be identified:

- It prepares the endometrium for pregnancy. Progesterone levels drop if pregnancy does not occur after ovulation.

- It inhibits contraction of smooth muscle in the uterus.

- It inhibits the development of new follicles.

Progesterone also plays key roles in pregnancy, labor, and after birth. During pregnancy the placenta secretes progesterone, taking over for the corpus luteum. Levels of progesterone are elevated during pregnancy, and they act to inhibit ovulation and lactation, but progesterone causes the growth of milk producing glands. A drop in progesterone levels may be involved in inducing labor and in the triggering of lactation.

In addition to its central role in pregnancy, progesterone plays other roles as well. It is known that the brain contains large number of progesterone receptors, and progesterone may be involved in memory and learning in the brain.

Aldosterone is a hormone derived directly from progesterone. This hormone is manufactured in the adrenal cortex, and is an example of a class of steroids we call *mineralocorticoids*. A mineralocorticoid is a hormone which is involved in the regulation of fluid volume and sodium in the body. In the case of aldosterone, three effects can be identified:

- It increases the reabsorption of sodium in the blood.

- It stimulates the release of potassium.

- It increases fluid volume.

From these three primary effects, it can be seen that the overall impact of aldosterone is a rise in blood pressure. If levels of aldosterone are too high in the body, high blood pressure and muscle cramps can result. *Renin* is an enzyme released by the kidneys in response to low blood volume or low sodium concentrations, and it stimulates the production of aldosterone. Pregnant women often have high levels of aldosterone.

The major *sex hormones* are also derived from progesterone. These include *testosterone,* which is synthesized primarily in the testes of males but also in the adrenal glands and ovaries of females. Testosterone is responsible for the so-called secondary sex characteristics in males, such as facial hair and deepening of the voice. It is derived directly from a 19 carbon hormone called *androstenedione.* In the female, secondary sexual characteristics are regulated by a hormone called *estradiol,* shown in Fig. 12-3, which is a derivative of testosterone. Estradiol is also produced in smaller amounts in males. Interestingly, estradiol plays a role in the regulation of gene transcription. It has an effect on bone and is also involved in the production of lipoproteins in the liver.

Cortisol is another hormone synthesized from progesterone that is produced in the adrenal cortex of the brain. It is classified as a *glucocorticoid,* meaning that it is a steroid which is involved in the metabolism of glucose. However, cortisol has wide and varied effects throughout the body. Cortisol may be involved in the regulation of the daily cycle of the body, as it helps restore homeostasis after periods of sleep. In the early morning, levels of cortisol are at their highest, while levels of cortisol are at their lowest about 3 hours after the onset of sleep. Cortisol also inhibits bone formation, has effects on the immune system, and increases sodium uptake. As a result cortisol increases blood pressure. Levels of cortisol are regulated in the body by two enzymes known as *enzyme 11-β hydroxysteroid dehydrogenase type I* (11-β HSD1) and *type II* (11-β HSD2). Local concentrations of cortisol are increased by the action of 11-β HSD1, which converts a biologically inactive compound called *cortisone* into cortisol. The reverse process, mediated by 11-β HSD2, converts cortisol back into cortisone decreasing cortisol levels.

Vitamin D

One other steroid hormone is of particular interest. This is *vitamin D*, a steroid that is obtained either from the diet or from sun exposure. It comes in two primary forms, called *vitamin D$_2$* (obtained from the diet) and *vitamin D$_3$* (*cholecalciferol*— obtained from sun exposure—synthesized in the skin from 7-dehydrocholesterol).

Vitamin D is not by itself useful to the body. It must be converted into a biologically active form by a two-step process that takes place in the liver and kidneys. Vitamin D$_3$ is far more effective than vitamin D$_2$ in the production of the biologically active steroid (may be up to 10 times as effective). In the case of vitamin D$_3$, the molecule is hydroxylated in a reaction catalyzed by *25-hydroxylase* yielding a molecule called *25-hydroxycholecalciferol.* The production of this intermediate molecule is not strongly regulated in the body, so levels of this chemical reflects levels of vitamin D consumption in the diet and sun exposure. The final step in synthesis of the biologically active form of vitamin D take place in the kidneys, where the conversion of 25-hydroxycholecalciferol to 1,25-dihydroxycholecalciferol is mediated by an enzyme

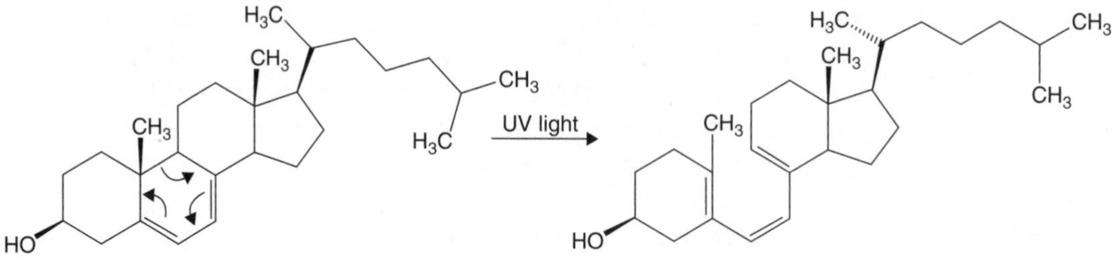

Figure 12-4 Exposure to UV light causes the molecule 7-dehydrocholesterol to transform into a molecule loosely called *pre-vitamin D₃*.

called *1-α-hydroxylase.* The first three steps in the synthesis of the biologically active form of vitamin D are illustrated in Figs. 12-4 to 12-6.

The major function of vitamin D in the body is the promotion of the intestinal absorption of calcium. It does this by promoting the manufacture of proteins in the cell that transport calcium from the intestine to the bloodstream. As such, vitamin D plays a key role in the maintenance of healthy bones. It plays a key role in several bone diseases. In *rickets,* a vitamin D deficiency in childhood leads to weakened and deformed bones. In adults vitamin D deficiency plays a role in osteomalacia and osteoporosis.

Recent research indicates that vitamin D plays an even wider role in the body. Vitamin D receptors exist throughout the body, including in the brain, gonads, heart, skin, prostate gland, and breasts. Vitamin D influences the growth and differentiation of cells, and may be implicated in cancer. In fact, vitamin D deficiency has been shown to correlate with increased risk for many types of cancer.

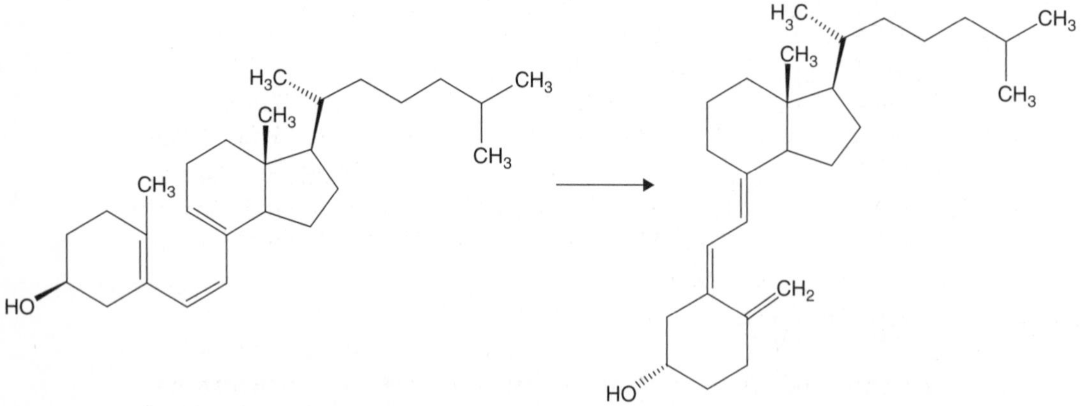

Figure 12-5 Pre-vitamin D₃ then spontaneously converts into vitamin D₃.

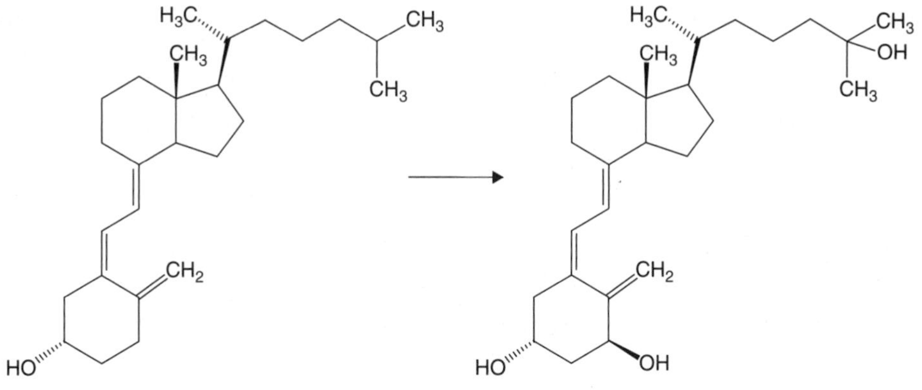

Figure 12-6   Conversion of vitamin D$_3$ into 25-hydroxycholecalciferol.

MECHANISM OF ACTION OF STEROID HORMONES

The way that hormones exert their effect on the cell is one of the most fascinating processes in the body. The primary method of action of steroid hormones is surprisingly to stimulate or inhibit the transcription of genes by the cell. Steroid hormones are lipids. As such they pass directly through the plasma membrane of the cell where they can bind to *intracellular receptors* which couple directly to nucleotide sequences called *hormone response elements* (HRES). Hormone receptors can be found in the cytosol or actually inside the nucleus of the cell.

An intracellular receptor is a polypeptide chain. On one end, we find a carboxyl group which marks the location of a *ligand-binding domain* to which the hormone binds. Upstream from the ligand-binding domain, there exists a *DNA-binding domain* which consists of amino acids that will bind the receptor to a specific DNA sequence. Finally, at the other end we find a region which activates transcription, this amino acid sequence is terminated by an amino group and so is known as the *amino terminus.*

When a hormone binds to a receptor, it causes conformational changes that give the receptor the ability to bind to DNA. We say that the hormone *activates* the receptor. The activated receptor then binds to a promoting region of the DNA specific to genes that are regulated by the hormone. After binding, the transcription of the gene is stimulated in most cases (hormones can also act to inhibit gene transcription).

PEPTIDE HORMONES

Several important hormones are short chains of amino acids called *peptide hormones.* These hormones are typically endocrine hormones which are secreted into the blood stream so that they can have influences far from their site of secretion. The *pituitary gland* plays a major role in the secretion of these substances. For example, *vasopressin* is a hormone secreted by the pituitary gland that has wide ranging effects, including reduction of urine volume, stimulation of water reabsorption by the kidneys, and stimulation of moderate vasoconstriction. These actions indicate that vasopressin acts to increase blood pressure, and it is also known as the *antidiuretic hormone.* A drop in blood pressure or blood volume will stimulate the release of vasopressin. This short chain polypeptide consists of 9 amino acids. Surprisingly, vasopressin may act in the brain to enhance pair bonding among mates in mammals.

Another example of a peptide hormone is *oxytocin,* which is a polypeptide made from 9 amino acids (Fig. 12-7). This hormone, which is made in the hypothalamus and secreted by the pituitary gland, also plays a wide range of roles in mammals. It is thought to be involved in social recognition and pair bonding, and is also secreted in large amounts in females during labor. It stimulates smooth muscle in the uterus, leading to contractions that deliver the fetus. After birth oxytocin levels in the cerebrospinal fluid of mothers is high, and it has been established that this hormone acts to promote bonding of the mother with her infant. Stimulation of the nipples of a nursing mother will cause the hypothalamus to release oxytocin. Oxytocin is also released in the brains of both sexes during orgasm, possibly enhancing pair bonding. Furthermore, it is believed to aid in the regulation of the circadian rhythm. During periods of high stress, the effects of oxytocin are reduced because oxytocin neurons in the brain are suppressed by *catecholamines,* which are neurotransmitters released by the adrenal gland during stress (see below).

Prolactin is a much longer polypeptide, consisting of about 195 amino acids together in a single chain. This hormone is also synthesized in the pituitary, but is also made by other cells such as cells in the immune system. Prolactin release is triggered in lactating females by infant stimulation of the nipples. The primary

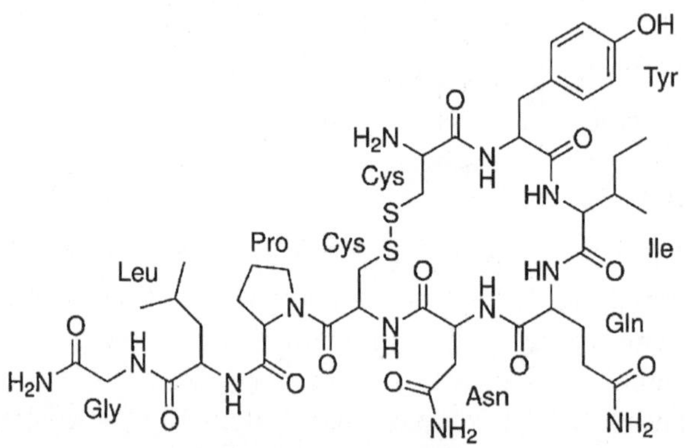

Figure 12-7 Oxytocin is a peptide hormone consisting of a chain of 9 amino acids.

function of the hormone is to stimulate development of the mammary glands and to trigger milk production. It has other actions as well; recent research indicates that during sexual arousal prolactin acts against dopamine to cause the so-called refractory period. In males, prolactin may cause impotence and loss of sexual desire in some cases.

Another important polypeptide hormone is *insulin,* which acts to stimulate the uptake of glucose by liver and muscle cells so that they can store it as glycogen. Insulin is synthesized in the pancreas by β-*cells,* which construct a single chain molecule called *proinsulin.* Enzymes excise a portion of the proinsulin molecule called the *C peptide,* producing the actual insulin molecule. When conditions warrant it, the β-cells will release insulin together with the c peptide into the blood stream via exocytosis.

The role of insulin in the body is well known, with its primary role being to control the uptake of glucose by liver and muscle cells.

Areas of Hormone Production

We have seen that certain areas of the body (such as the adrenal and pituitary glands) are involved in hormone production. Hormones are actually produced in many areas throughout the body, which we now summarize. Due to limited space, this list is necessarily incomplete, but is provided to give the reader an idea of the wide range of organs involved in hormone production and release.

ADRENAL CORTEX

The adrenal cortex is involved in the production of two types of steroid hormones, glucocorticoids which affect metabolism and decrease inflammation, and mineralocorticoids which act to maintain salt and water balances.

ADRENAL MEDULLA

The *medulla* or core of the adrenal gland is responsible for the production of two important hormones which are amino acid derivatives. These include *epinephrine,* which causes the contraction of the smooth muscles, increases hear rate and blood pressure, and stimulates glycogenolysis in liver and muscle cells. The adrenal medulla also synthesizes norepinephrine which stimulates arteriole contraction and decreases peripheral circulation among other effects. We will discuss both hormones in more detail when we discuss neurotransmitters, below.

INTESTINE

Hormones are also produced in the intestines. Two examples include *cholecystokinin* (CCK), which is a polypeptide hormone that causes the pancreas to release digestive enzymes. In addition, it also stimulates the gall bladder causing it to empty.

LIVER

The liver is involved in the synthesis and secretion of several hormones. Insulin-like growth factor-1 (IGF-1) or somatomedin is a polypeptide hormone which stimulates cartilage growth. Cells in the bone marrow are also affected by this hormone. It can bind to cells and trigger mitosis. The release of IGF-1 is triggered by the binding of growth hormone to liver cells.

Another important hormone produced by the liver is *thrombopoietin,* is a 332 amino acid chain polypeptide. This hormone acts in the bone marrow, where it causes cells to differentiate into *megakaryocytes* which generate blood platelets.

PLACENTA

We have already seen that the steroid progestin hormones are produced in the placenta, where they play a role in regulation of the menstrual cycle and pregnancy. The placenta also produces a polypeptide hormone called *chorionic gonadotropin* that acts to cause the release of progesterone.

STOMACH

The stomach also produces several hormones, which are peptides. An example is *gastrin,* which triggers cells in the stomach to secrete gastric juice.

THYROID

The thyroid gland produces several vital hormones. The polypeptide *calcitonin,* which consists of a chain of 32 amino acids, acts to lower the amount of calcium in the bloodstream. It does this by inhibiting calcium uptake by the intestines. The thyroid also produces two important hormones which are amino acid derivatives, *triiodothyronine* (T3) and *thyroxine* (T4). These hormones act to control the rate of metabolic processes, with T3 acting as a metabolic stimulator.

NEUROTRANSMITTERS

The nervous system is composed of cells called *neurons* which are linked together by structures called *synapses* which enable the cells to "talk" to each other. A synapse actually contains a small gap between the cells called a *synaptic cleft,* which is a gap on the order of 20 nm wide. Nerve cells can be viewed as having a treelike structure, with branches at the top connected to the cell body containing the nucleus, a trunk like structure called the *axon* and a system of "roots" at the bottom which connect to the branches of another nearby neuron with which it can communicate. A nerve cell works to transmit information via an electrical signal that travels down the axon. However, the existence of the gap (synaptic cleft) requires nerve cells to use chemical messengers to transmit the signal to the next neuron. These chemical messengers are waiting in vesicles that are released into the synaptic cleft via exocytosis. They diffuse across the gap and bind to receptors on the next neuron. We call the chemical messengers *neurotransmitters,* which are chemicals that serve to relay a signal from one neuron to another. This action serves to alter the behavior of neighboring neurons, causing them to "fire" or inhibiting them from doing so. The type of neurotransmitter released by a cell depends on the cell type and its location in the brain or peripheral nervous system.

There are several different classes of neurotransmitters. Several are *monoamines,* and include the famous chemicals *dopamine, serotonin,* and *norepinephrine* which are involved in the regulation of mood and emotion, among other functions. Polypeptides can also act as neurotransmitters; we have already met one in oxytocin. There are also smaller molecules which consist of a few amino acids such as *γ-aminobutyric acid* (GABA), an amino acid derivative which acts as a chemical messenger between neurons.

Many neurotransmitters also act as hormones in the bloodstream. However, they are not able to enter the brain this way because the brain is protected by a type of filter scientists call the *blood brain barrier.* The existence of this filtration system has important consequences in treating neurological disorders, because therapeutic agents must be administered as precursor molecules which can pass the blood brain barrier. Once inside the brain, they are processed into the desired chemical.

We now consider a few neurotransmitters in detail.

Epinephrine and Norepinephrine

Two important chemicals which act as neurotransmitters and hormones are epinephrine and norepinephrine, (also known by their older names adrenaline and noradrenaline) which are produced in the adrenal medulla. The base compound used in the synthesis of these two chemicals is the amino acid tyrosine. Norepinephrine, shown in Fig. 12-8, is an 8-carbon compound with the formula

$$C_8H_{11}NO_3$$

One of the primary roles played by norepinephrine is in the production of the *flight or fight response.* In short, given a stimulation which may be threatening, it

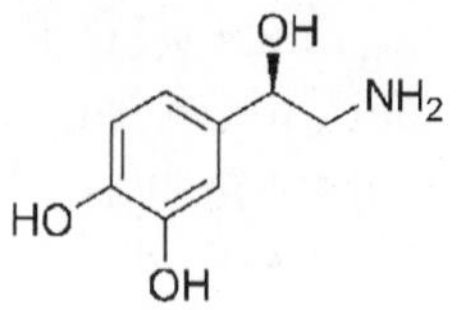

Figure 12-8 Norepinephrine is an important neurotransmitter.

helps ready the body for action. A threatening stimulus will cause the release of large amounts of norepinephrine which act through the sympathetic nervous system (SNS) to

- Cause the release of glucose.
- Increase the heart rate.
- Stimulate the skeletal muscles.

For these reasons this compound is known as a *stress hormone.* The activation of this response takes place in the brain stem, and so is automatic. In the bloodstream, this chemical acts to cause vasoconstriction, an effect which also readies the body for physical action.

Norepinephrine is also a key neurotransmitter in the brain. It plays several roles, including

- The regulation of mood and emotion.
- Maintenance of attention and focus.
- Depression.

The role of norepinephrine in clinical depression is well established. Compounds called *reuptake inhibitors,* which delay norepinephrine from being taken up from the synaptic cleft and hence prolong the action of the chemical, enhance the mood of depressed patients.

Tyrosene is oxidized by tyrosine hydroxylase into a compound called L-*dopa,* which is the rate-limiting step in the production of norepinephrine. L-dopa then undergoes decarboxylation by pyridoxal phosphate and *dopa* decarboxylase to produce *dopamine,* which is itself an important neurotransmitter. An enzyme called *dopamine* β-*hydroxylase* then transforms dopamine into norepinephrine.

When an animal is faced with a threat, the related compound epinephrine is also released, acting to trigger the flight or fight response. Using older or common terminology, epinephrine is sometimes called adrenaline. After norepinephrine has been produced, epinephrine can be synthesized in a single step. This is done with the aid of an enzyme called *phenylethanolamine N-methyltransferase,* which acts to add a methyl group to norepinephrine. The result is epinephrine which is illustrated in Fig. 12-9.

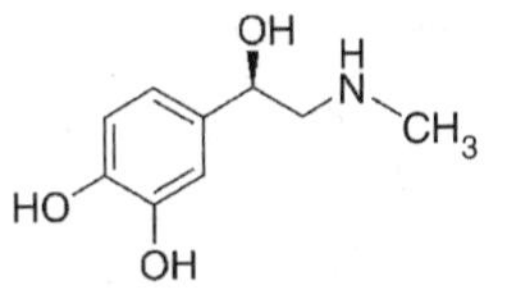

Figure 12-9   Epinephrine, an important neurotransmitter and hormone involved in stimulation of the sympathetic nervous system.

The catecholamines, including epinephrine and norepinephrine, act on cells through receptors on the cell membrane called *adrenergic receptors*. There are two major classes of adrenergic receptors, which can be denoted as α- and β-receptors. When epinephrine binds to β-adrenoreceptors, the following effects take place in the body via the activation of *adenylate cyclase:*

- Glycolysis is stimulated in skeletal muscle cells.
- Gluconeogenesis is stimulated in the liver.
- Lipolysis is stimulated in adipose tissue.
- Smooth muscles in blood vessels supplying skeletal muscles relax.
- Heart rate is increased.
- Smooth muscles in the bronchial tubes relax.

We see that the *same* receptors together with the *same* hormone (epinephrine) serves to achieve one goal—preparing the body for the flight or fight response. When epinephrine binds to α-adrenergic receptors, the following effects are noted:

- Smooth muscle contraction in blood vessels supplying the peripheral organs is stimulated.
- Smooth muscle *relaxation* in the lungs and gastrointestinal tract.
- Blood platelet aggregation is stimulated, preparing the body for clotting.

The α-adrenergic receptors can be further classified into α_1 and α_2 receptors. The α_1 receptors act via a process called the *phosphoinositide cascade,* while the α_2 receptors act by inhibition of an enzyme called *adenylate cyclase.*

In medicine, the action of a receptor is often characterized by the use of *agonists* and *antagonists.* An agonist is a substance that binds to a hormone receptor and induces a hormone response. In contrast, an antagonist binds a hormone receptor, but *does not* induce a hormone response. As a result the antagonist prevents the hormone from binding to a cell and blocks the action of the hormone. Agonists and antagonists can often be used as therapeutic agents. For example, a chemical called *propranolol* blocks α-adrenoreceptors and hence acts as an antagonist. By noting

the action of epinephrine on α-adrenoreceptors described above, the reader should not be surprised to learn that this compound can be used to treat high blood pressure in some cases.

Now let's consider how epinephrine acts at the cell membrane level. A sudden shock which stimulates the flight or fight response (such as someone cutting you off in traffic) results in the rapid release of epinephrine into the blood stream. This stimulates the liver to degrade glycogen making more glucose available to the body, and causes the production of glucose 6-phosphate in the skeletal muscle cells, giving them the energy they need to take action. These processes are mediated by cAMP.

At the cell membrane, epinephrine (the *first messenger*) binds to a receptor in the phospholipid bilayer which causes in conformational change in membrane bound *adenylate cyclase.* When activated, ATP can be converted into cAMP which acts as a *second messenger* inside the cell.

Dopamine

Another important catecholamine which plays a role in the brain and in the sympathetic nervous system is dopamine. Improper levels of this relatively simple molecule, illustrated in Fig. 12-10, are implicated in attention deficit disorder, schizophrenia, and Parkinson's disease.

In the brain, dopamine is produced in the hypothalamus and a region called the *substantia nigra.* It serves as an important mediator of behavior, perception, and memory, and is also involved in the regulation of motor activity and attention and learning.

In regard to mental function, excess and reduced dopamine levels lead to very different effects. This is particularly important in the *frontal lobes,* which are involved in the regulation of information flow to other parts of the brain. High dopamine levels in the brain lead to schizophrenia, a disease characterized by flawed thought processes, flawed perceptions, and hallucinations. Sufferers of this disease are often paranoid, believing that government agents or others are "after" them, or they have imagined accomplishments such as believing they invented hairspray or have helped aliens design spacecraft. Moreover, they usually hear voices or see people who are not really there.

In the brain there are five types of dopamine receptors conveniently denoted D1 to D5. A medication called *phenothiazine* acts to block D2 receptors on neurons, mitigating the effects of increased dopamine levels in the brains of schizophrenics.

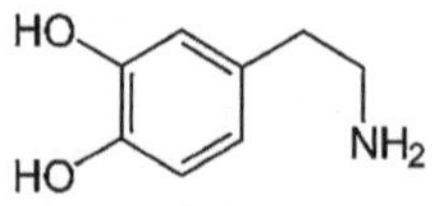

Figure 12-10   The humble dopamine molecule is vital for healthy brain function.

Clinical manifestations of the disease such as auditory and visual hallucinations are reduced by the drug.

In contrast, low dopamine levels in the prefrontal cortex can lead to a condition called *attention deficit disorder* (ADD), a condition which is characterized by problems with focus and sitting still (such as in a classroom). Medications which increase dopamine levels (such as *Ritalin*) mediate the clinical symptoms of this condition. Interestingly, in animal models, the administration of antipsychotics such as Haldol which act to reduce dopamine levels can induce an ADD-like response.

Dopamine is also an important neurotransmitter among neurons involved in motor control. In particular, the death of cells in the substantia nigra region of the brain, cells which influence the *basal ganglia* which is involved in the regulation of smooth muscle movement, leads to Parkinson disease. This condition can be mediated by the increasing dopamine levels in the brain. Dopamine cannot cross the blood-brain barrier, so the smaller precursor molecule L-DOPA is administered.

In addition to mediating mood, behavior, and motor activity, dopamine acts on the sympathetic nervous system to increase heart rate and blood pressure. Furthermore, it acts as a hormone to inhibit the release of prolactin. It is also believed to be involved in sexual arousal.

Serotonin

One of the most noted neurotransmitters in mental health is *serotonin*. This important compound is involved in mood regulation and the inhibition of anger and anxiety. Low levels of serotonin can be a key factor in clinical depression.

Serotonin, shown in Fig. 12-11, is synthesized from *L-tryptophan*. The synthesis of serotonin can be described as a two step process. In the first step, an enzyme called *tryptophan hydroxylase* (TPH) catalyzes the transformation of tryptophan into 5-hydroxy-L-tryptophan. Then, an amino acid decarboxylase converts this compound into serotonin. It is released into the brain by neurons in a region called the *raphe nuclei*. Tryptophan can cross the blood brain barrier, so tryptophan ingested in the diet will result in the production of serotonin. Serotonin is also involved in the regulation of sleep, and turkey contains large amounts of tryptophan which may explain why holiday meals often leave you tired.

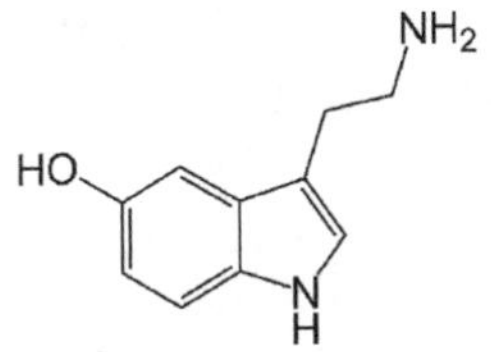

Figure 12-11   Serotonin is an important neurotransmitter vital for the healthy
regulation of mood.

Low serotonin levels are associated with clinical depression. Two ways that we can deal with low serotonin levels in the brain are by the administration of *monoamine oxidase inhibitors* and reuptake inhibitors. Monoamine oxidase is an enzyme which degrades serotonin in the synaptic cleft, thereby limiting its action. In a patient with low serotonin levels, it is desirable to maximize the impact of the serotonin present in the brain. This can be accomplished by the administration of drugs called monoamine oxidase or MAO inhibitors, which inhibit the action of this enzyme in the synaptic cleft, preventing the breakdown of serotonin. Clinically, this has the effect of improving mood but these drugs have many dangerous side effects.

Reuptake inhibitors, which were briefly mentioned in our discussion of norepinephrine, inhibit the reuptake of serotonin in a synapse allowing it to act longer. These chemicals are called *tricyclic antidepressants.*

GABA

GABA is one of the most ubiquitous neurotransmitters in the brain. In fact, 30% of the brains synapses are GABA synapses. GABA is an inhibitory neurotransmitter, acting to prevent the postsynaptic neuron (the receiver of GABA) from firing. It does so by binding to the target neuron, causing ion channels in the cell to open. This allows positively charged potassium ions and negatively charged chlorine ions to cross the cell membrane such that an overall negative electrical potential is set up along the axis of the neuron, effectively shutting it off until it receives some other stimulation. GABA is synthesized from *glutamate* using an enzyme called L-*glutamic acid decarboxylase.* Interestingly, glutamate also acts as a neurotransmitter, but one which is excitatory. The principal role of glutamate is to strengthen synaptic connections (so it may be involved in memory formation).

Acetylcholine

The final neurotransmitter we will examine is *acetylcholine.* You may be familiar with the role that this compound plays in the activation of voluntary muscle movement. In addition to being important in the peripheral nervous system, acetylcholine is found in large quantities in the brain.

Acetylcholine is manufactured from choline and acetyl coenzyme A using the enzyme *choline acetyl transferase.* It has the chemical formula

$$CH_3COOCH_2CH_2N+(CH_3)_3$$

In the body, acetylcholine plays an important role in the contraction of voluntary muscles. Normally, after release by neurons which stimulate the muscles, it is broken down by an enzyme called *acetylcholinesterase.* Early in the previous century some biologists figured out that they could create substances called *nerve gases* that would kill soldiers on the battlefield by interfering with this process. In short, nerve gases act by preventing the degradation of acetylcholine causing muscle spasms which lead to rapid death.

In nature, several substances also interfere with the normal action and breakdown of this neurotransmitter. Botulism toxin inhibits acetylcholine release, leading to a type of paralysis. In contrast, black widow venom (called *latrotoxin*) causes excessive release of acetylcholine (along with norepinephrine and GABA), preventing muscle relaxation. This can cause a tetanus-like effect leading to severe muscle cramping and pain, with the muscles of the abdominal region particularly affected, leading to breathing problems.

In the brain, acetylcholine is an important neurotransmitter involved in learning and memory. Alzheimer disease is caused by the deaths of large number of neurons that use and synthesize acetylcholine. Hence low acetylcholine levels are associated with Alzheimer disease.

Another disease involving this neurotransmitter is *myasthenia gravis,* which is an autoimmune disease involving the destruction of acetylcholine receptors. Symptoms of this disease include muscle weakness and fatigue.

Summary

In order to function as a unified whole, a multicellular organism requires a communication system to transmit signals from one body area to another. The chemical messengers which play this role are called hormones. This process is taken to another level in the brain, where chemical messengers called neurotransmitters act as mediators in communication between nerve cells.

Quiz

1. Endocrine hormones

    (a) are released by ductless glands and act on nearby cells via diffusion.

    (b) are released into the bloodstream for action on receptor cells distant from the source.

    (c) are released only by the adrenal glands.

    (d) bind to special proteins which enable them to act on the DNA.

2. Vitamin D

   (a) is not a hormone, it is an element consumed in the diet.

   (b) is a peptide hormone that must be processed in the body into a biologically active form.

   (c) is a steroid hormone which must be processed in the body into a biologically active form.

   (d) is a steroid hormone which does not need to be processed by the body into a biologically active form.

3. Cholesterol

   (a) is a precursor to most steroid hormones. The first step in the synthesis of a hormone involved the cleavage of a 6-carbon residue from cholesterol.

   (b) is not involved in the synthesis of most steroid hormones. It is an undesirable component of the diet.

   (c) is a precursor to most steroid hormones. The first step in the synthesis of a hormone involved the addition of a 6-carbon residue from cholesterol.

   (d) is a precursor to all steroid hormones.

4. A pregnane is a steroid hormone with

   (a) 19 carbon atoms.

   (b) 17 carbon atoms.

   (c) 18 carbon atoms.

   (d) 21 carbon atoms.

5. Cortisol can be characterized as a hormone which

   (a) increases blood pressure.

   (b) increases vasoconstriction, but has minimal effect on blood pressure.

   (c) reduces blood pressure.

   (d) is only involved in stressful effects on the immune system.

6. High levels of dopamine are implicated in

   (a) Parkinson disease

   (b) attention deficit disorder.

   (c) schizophrenia.

   (d) Alzheimer disease.

7. An Antagonist is best described by which of the following?

   (a) It can bind to a hormone receptor, preventing a hormone from binding to a cell and producing other, unexpected effects.

   (b) It can bind to a hormone receptor, preventing a hormone from binding to a cell and thereby the effect of the hormone is absent.

   (c) It is a compound which blocks the action of an agonist.

   (d) It is a compound which acts to degrade a hormone.

8. Which of the following best describes the synthesis of catecholamines?

   (a) Epinephrine is a precursor to norepinephrine.

   (b) Norepinephrine is a precursor to epinephrine.

   (c) Dopamine is the final product in catecholamine synthesis.

   (d) Dopamine is made into epinephrine, but not norepinephrine.

9. Norepinephrine acts on the sympathetic nervous system

   (a) to increase heart rate and cause the release of glucose.

   (b) to increase heart rate and inhibit the release of glucose.

   (c) relax all smooth muscles.

   (d) contract all smooth muscles.

10. Nerve gases act by

   (a) blocking acetylcholine receptors.

   (b) inhibiting acetylcholine release.

   (c) inhibiting the action of acetylcholinesterase, preventing breakdown of acetylcholine.

   (d) enhancing the action of acetylcholinesterase, causing rapid breakdown of acetylcholine.

CHAPTER 13

Specialty Functions

PART A Hemoglobin

Introduction

The truly astounding capabilities of biomolecules can be best appreciated by analyzing two processes essential to life—the delivery of oxygen to multicellular tissue by hemoglobin and the fixation of energy through the process of photosynthesis. These processes are accomplished by eerily similar molecules, as you will see. Needless to say, dear reader, without these two functions neither you nor I nor anybody else would be here.

We will start with hemoglobin. Obviously, life forms would be very limited if we all had to depend on the dissolution of oxygen into blood, that is, were there not a transport system for oxygen. For most of us, oxygen has insufficient solubility in blood, unless you happen to be an arctic ice fish. For the ice fish, the cold temperatures of the artic environment increase the solubility of oxygen in the fish's blood. This, coupled with a pretty sluggish fish, makes survival possible without an oxygen transport molecule. For the rest of us, we must have a molecule that reversibly binds oxygen from the atmosphere, retains it through most of its journey through the blood stream and releases oxygen in the tissues. There are several similar molecules in living systems that do the trick, but hemoglobin is the best.

To understand the phenomena of oxygen transport, you need to know the following:

- The molecule structure of oxygen transport molecules.
- When and why oxygen is released.
- Factors that modulate oxygen transport and oxygen release.
- Other functions of hemoglobin.
- Chemicals that impede hemoglobin functions—poisons.
- Disease conditions resulting from abnormal hemoglobin.

The Molecular Structure of Oxygen Transport Molecules

The availability of oxygen at a cellular level makes possible oxidative metabolism. This allows quite an energy boost because we now have an order of magnitude jump in energy availability over nonoxidative metabolism. The biomolecules engineered to carry oxygen to the tissues includes some exotic types found in insects and other nonmammalian species such as hemocyanin and hemerythrin. In this discussion, we aim to understand the basic mechanisms of oxygen transport and will concentrate on *myoglobin* and *hemoglobin*. The basic organization and principles are the same.

Myoglobin is a single polypeptide chain and is found predominantly in the muscle tissue.

The polypeptide chain of myoglobin is arranged as a series of α helices as shown in Fig.13-1. This configuration is common with water-soluble globular proteins. As you would expect, the interior of the helices is occupied by hydrophobic amino acid side chains and out of the exterior protrude hydrophilic amino acid side chains.

The α-helix structure forms a cage to contain the functional portion of the myoglobin protein, the *heme* group. The heme is suspended within the polypeptide cage. See Fig. 13-2.

Let's review some terms in case it has been a while since you have heard them. A *conjugated protein* contains structures that are not amino acids. Such non-amino acid structures are called *prosthetic groups*. Myoglobin is a conjugated protein because of the presence of the heme prosthetic groups. The protein portion of a

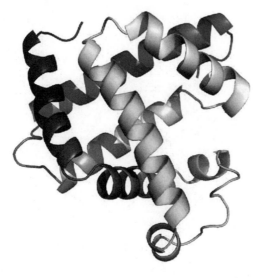

Figure 13-1   Myoglobin polypeptide chain.
(*Image courtesy of Wikipedia Commons*)

conjugated protein is called an *apoprotein*. Hemes contain iron. This heme holds four *pyrrole* subunits interconnected via their α-carbon atoms by methine bridges (=CH-). A pyrrole is a five-membered ring containing nitrogen. The overall organic net holding the iron is called a *protoporphyrin*. So you have a prosthetic group consisting of a protoporphyrin, composed of four pyrrole rings, that holds an iron, making it a heme. See Fig. 13-3.

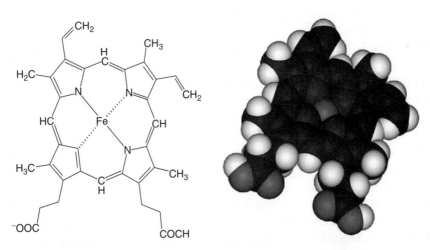

Figure 13-2   The heme group of myoglobin and hemoglobin.

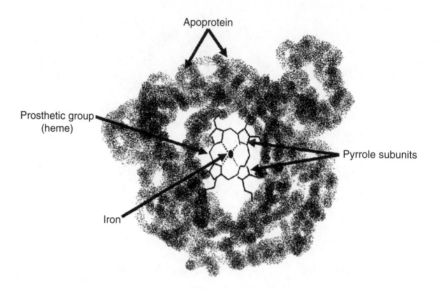

Figure 13-3 Depiction of a subunit of hemoglobin.

The Fe is bound on four sides by covalent bridges with nitrogen. The iron in the ferrous state (Fe^{2+}) has two more binding sites. These are called the fifth and sixth coordination sites, for obvious reasons, and can bind suitable substrates above and below the plane of the heme. One bonds a histidine from the surrounding protein— called the proximal histidine to distinguish it from other histidines in the protein. The coordination site with the proximal histidine further stabilizes the iron. The remaining coordination site bonds the oxygen. The oxygen bond is further stabilized by a hydrogen bond between the oxygen and a histidine within the protein structure called the distal histidine.

Here's the rub. The unoxidized iron is in the ferrous state (Fe^{2+}). You don't want the oxygen to oxidize the iron to the ferric state (Fe^{3+}). If this were to happen, an unfortunate event would follow because Fe^{3+} does not bind oxygen! The oxygen would be released from the transport molecule *and* it would be released as a highly reactive free radical. The oxidation of the iron is prevented by the bulky surrounding protein structure. The protein functions to partially shield the iron from oxygen. As a result, an electron is only partially transferred from the iron to the oxygen.

Hemoglobin is much bigger than myoglobin. It is comprised of four polypeptide chains each bearing a heme group, and is found in the red blood cells. The two proteins, myoglobin and hemoglobin are obviously related. The four polypeptide chains of hemoglobin are of two types, called the α chains and the β chains. Each hemoglobin protein contains two α and two β chains. The two types of chains are quite similar, differing only by a few amino acids, and myoglobin is similar to both.

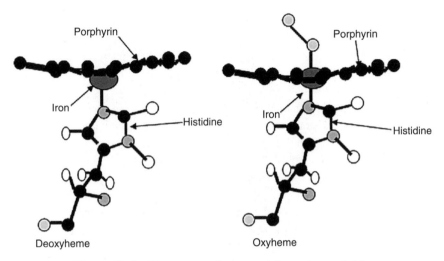

Figure 13-4   Heme group in oxy and deoxy hemoglobin.

It very well may be that hemoglobin evolved from myoglobin. The arrangement of hemoglobin is such that the α and β chains pair, with one α and one β chain closely associated in a dimer. One dimer is called $\alpha_1\beta_1$ and the other is $\alpha_2\beta_2$.

The configuration of hemoglobin changes slightly when oxygen is bound. The *deoxyhemoglobin*, with no oxygen bound, is called "T" for *tense state*. Oxyhemoglobin is formed when oxygen is bound and the molecule relaxes, allowing the Fe to move closer into the protoporphyrin plane. The state is referred to as "R" for *relaxed state*. (Not all biochemical names are esoteric.)

In the deoxyhemoglobin, the molecules are positioned relatively tightly. In fact, the iron atom doesn't quite fit within the plane formed by the protoporphyrin and sits a little below. After oxygen is bound, the molecular structure relaxes a bit, and the iron moves closer into the protoporphyrin plane. See Fig. 13-4 for a depiction of oxy and deoxy hemoglobin.

Release of Oxygen

Oxygen is present in the biofluids as a dissolved gas. The quantity of a gas dissolved in a liquid can be estimated by the pressure exerted by the gas at the liquid surface, known as the *partial pressure*. For oxygen, the symbol for partial pressure is P_{O_2}. This is also known as *oxygen tension*. Pressure is expressed in torrs. One torr is the pressure required to raise a column of mercury one millimeter. Hemoglobin picks up or releases oxygen in response to the amount of oxygen in the biofluids, that is, in proportion to P_{O_2}.

The degree to which the available hemoglobin has taken up oxygen is measured as fractional saturation, or "Y". For example, if Y is 0.5, half of the binding sites of hemoglobin are occupied. The degree of fractional saturation is one measure of how much oxygen can be made available to the tissues. Obviously, the amount of hemoglobin in the blood also limits oxygen availability.

Hemoglobin is remarkably good at binding oxygen. Hemoglobin can reach a fractional saturation of nearly 0.9 whereas myoglobin can achieve only about 0.1. The reason is that oxygen binding in hemoglobin exhibits *cooperativity*. Each hemoglobin molecule can bind 4 molecules of oxygen. The binding occurs with increasing ease with each oxygen that binds. The second oxygen binds three times more easily than the first. The fourth oxygen binds 40 times more easily than the first. As a result, the oxygen binds at an increasing rate as the presence of oxygen increases. In a plot of fractional saturation against P_{O_2}, a sigmoid curve is generated wherein the oxygen binds slowly at first and then is accumulated at an increasing rate until the hemoglobin nears saturation. This plot is also referred to as *the oxyhemoglobin dissociation curve*. See Fig. 13-5.

The cooperativity behavior is attributable to the transition from the "T" to the "R" configuration. When the first oxygen binds, it causes a change in the tertiary structure of the entire subunit. Conformational changes at the subunit surface lead to a new set of binding interactions between adjacent subunits, affecting ionic interactions, hydrogen bonds, and hydrophobic interactions. The changes in subunit interaction affect the heme-binding pocket of a second deoxy subunit and result in easier access of oxygen to the iron atom of the second heme. The effect is cumulative

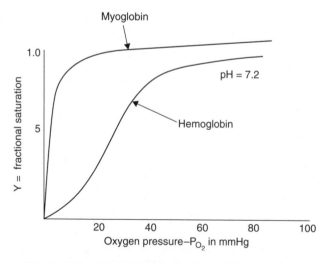

Figure13-5 Fractional saturation of hemoglobin and myoglobin with changes in P_{O_2}.

so that when three subunits are in the relaxed configuration, having bound oxygen, the fourth oxygen binds with extraordinary ease

The hard part of oxygen transport is knowing when to let go. The idea is to hang on to the oxygen until the red blood cell (RBC) is in the capillaries, where released oxygen will diffuse into the tissues and will be available for oxidative metabolism. Oxygen pressure in serum drops as the blood moves from the lungs. In the lungs, the P_{O_2} is about 100 torr. In the tissues the P_{O_2} drops to around 20 torr. The P_{O_2} may be much lower in cells that are consuming high amounts of oxygen to generate energy, such as exercising muscles. Review the curve exhibited in Fig. 13-5 and you can see that hemoglobin absorbs oxygen very well at a P_{O_2} of 100 torr where there is sufficient O_2 to overcome the inhibitory nature of the T state. The majority of this oxygen is released at a P_{O_2} of 20 to 40 torr, the P_{O_2} found in the capillaries. The lower P_{O_2} favors the T state.

Myoglobin cannot undergo cooperative binding because it has only one binding site and it binds oxygen much more firmly than hemoglobin. The majority of the oxygen bound by myoglobin is released in pressure ranges below 20 torr. Therefore, the myoglobin stores oxygen under conditions where hemoglobin is depleted of oxygen. Historically, myoglobin was described as enabling a "bucket brigade" transfer of oxygen. Oxygen theoretically would pass from hemoglobin in the blood to myoglobin in the muscle, to be released under conditions of oxygen deprivation such as heavy exercise. However, recent experiments with mice that were engineered to fail to produce myoglobin have raised some doubts. These mice showed no impairment, even with exercise. Myoglobin may have other functions related to the regulation of NO, such as regulating capillary pressure, as discussed below.

Factors That Modulate Oxygen Transport and Release

There are a number of factors that influence the strength of oxygen binding. Factors that strengthen oxygen binding cause a *left shift* to the dissociation curve. Factors that lessen the strength of oxygen binding cause a *right shift* to the dissociation curve. Remember this—*A right shift of the dissociation curve increases oxygen delivery to the tissues.*

Factors that decrease the affinity of hemoglobin for oxygen do so by stabilizing the T state. This is because the T state binds oxygen very poorly. Among these factors are pH and *2,3-biphosphoglycerate* (2,3-BPG).

pH

Note that hemoglobin contains amino acids that can accept or donate protons. Hemoglobin is a weak acid, tending to give up protons at physiological pH. The loading and unloading of oxygen by hemoglobin influences pH, because, during the O_2 binding-induced alteration from the T form to the R form, several amino acid side groups on the surface of hemoglobin subunits will dissociate protons.

$$Hb(O_2)_nH_x + O_2 = Hb(O_2)_{n+1} + xH^+$$

An increase in the hydrogen ion concentration—that is, a decrease in pH—decreases the affinity of hemoglobin for oxygen. This means that hemoglobin will release the oxygen more readily. The fractional saturation of hemoglobin is less at a given P_{O_2} so the curve of fractional saturation versus P_{O_2} is shifted to the right. The decreased affinity of hemoglobin for oxygen with decreasing pH is called the *Bohr effect*. See Fig. 13-6.

The lower pH stabilizes the T configuration by enabling the formation of ionic, or salt, bridges between amino acids whose pK_a favors dissociation during acid conditions.

The *Bohr effect* is less in the lungs, thankfully, and is more active in exercising muscles, where the buildup of CO_2 and of lactic acid serve to lower the pH.

Conditions inside the red blood cells (RBCs) help lower the pH in the capillaries, where CO_2 is evolved by metabolizing tissues. RBCs contain an enzyme *carbonic anhydrase*, which catalyzes the conversion of CO_2 to carbonic acid as follows:

$$CO_2 + H_2O = HCO_2^- + H^+$$

The HCO_3^- diffuses out of the RBC and is transported in the serum to the lungs. The majority of CO_2 transport to the lung occurs as HCO_3^-. As the HCO_3^- diffuses

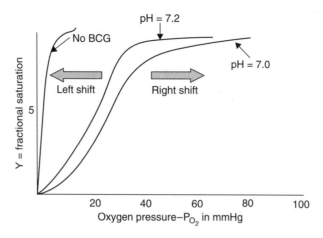

Figure 13-6 Effect of pH and BCG on the oxyhemoglobin dissociation curve.

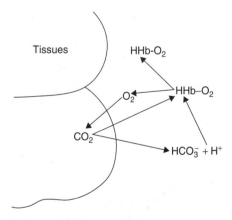

Figure 13-7   Formation of
carbamates in the presence of CO_2.

out of the RBC, Cl^- diffuses in to balance the charge. The chlorine also acts to
stabilize the T state of the hemoglobin, thereby further enabling O_2 release. The H^+
is taken up by the hemoglobin, which acts as a buffer. Upon return to the lungs, the
hemoglobin release the H^+ as it takes up oxygen. The H^+ reacts with HCO_3^-, forming
carbon dioxide and water. The system is designed to optimize both oxygen release
and CO_2 transport in the tissues.

   CO_2 stabilizes the T state even when the pH is held constant in a manner that also
enables CO_2 transport. Some of the CO_2 reacts with the N-terminal amine group to
form a carbamate. See Fig. 13-7.

   The negatively charged carbamate groups participate in ionic bonds and thereby
stabilize the T state.

   Actually, most of the CO_2 in the blood is transported in the form of bicarbonate in
the serum. As the bicarbonate is formed, the pH of the blood is lowered and the
release of oxygen is enhanced. See Fig. 13-8. Note that the system optimizes oxygen
release at the level of capillaries, where the pH is lowest and CO_2 concentration is
highest. It also enables CO_2 release in the lung, when O_2 concentration is high.

Figure 13-8   Interaction of hemoglobin
and CO_2 in gas transport.

2,3-BPG

The only reason that hemoglobin releases oxygen in the manner shown in Fig. 13-6 is because of the presence of the small biomolecule 2,3-BPG, which stabilizes the T configuration. The T configuration of hemoglobin is much less stable than the R configuration. Without the 2,3-BPG, the hemoglobin tends to stay in the R state and binds oxygen strongly. Under these conditions, oxygen is not released at the partial pressures found in the tissues.

The 2,3-BPG sits in a pocket in the middle of the tetramer, a pocket that exists only when oxygen is not bound, that is, in the T state. Bonds between the 2,3-BPG and the surrounding amino acids help to stabilize the T state. In the conversion to the R state, the pocket disappears and the 2,3-BPG pops out.

The concentration of 2,3-BPG increases under conditions of oxygen deprivation. For example, adaptation to high altitude is due in part to increased 2,3-BPG in the RBCs where it causes a right shift of the deoxyhemoglobin dissociation curve. A right shift means enhanced oxygen delivery to the tissues. 2,3-BPG increases in other situations where oxygen tension decreases, such as in individuals with sleep apnea. See Fig. 13-9. Changes in 2,3-BPG concentration can occur very rapidly and enable quick response to environmental changes.

When 2,3-BPG was first discovered, its action was part of the motivation for professional athletes to enhance their performance by a practice known as blood-packing. These individuals would have blood withdrawn and stored prior to an athletic event. Immediately before the event, they would have a transfusion of this blood, increasing their concentration of hemoglobin, their hematocrit, and, theoretically, their oxygen availability. If the stored blood was collected after acclimation to high

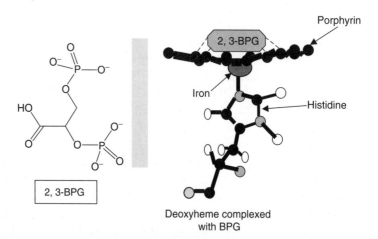

Deoxyheme complexed
with BPG

Figure 13-9   2,3-BPG

altitude, the presence of 2,3-BPG would be increased, resulting in decreased affinity for oxygen and improved oxygen delivery to the tissues. This practice is not only illegal but dangerous and is part of the screening protocol for many athletic events.

The effects of both hydrogen ions and 2,3-BPG on oxygen are examples of *allosteric regulation* of protein function. Allosteric effector molecules bind the molecule at a site other than the protein's active site and exert their effect by changing the confirmation of the molecule (see Chap. 9).

FETAL HEMOGLOBIN

Developing fetuses possess a distinct form of hemoglobin known as *fetal hemoglobin*. Fetal hemoglobin is composed of two α helices and two γ chains. The γ chains are similar to β chains except for key amino acid substitutions in sites which bind 2,3-BPG. Specifically, serine is substituted for histidine, removing two binding sites. As a result, 2,3-BPG has less affinity for fetal hemoglobin and fetal hemoglobin has higher affinity for oxygen. This higher affinity allows the fetal hemoglobin to draw oxygen from maternal red blood cells under condition of low oxygen pressure.

Other Functions of Hemoglobin

Hemoglobin is a beautiful control system, in that it not only fuels the energy mechanisms of the cell, but functions to deport waste products and in other transport schemes. The discussion above deals with the role of hemoglobin in the transport of CO_2, Cl^-, and protons to the lungs. In addition, hemoglobin, by binding and releasing nitric oxide (NO), plays a key role in regulating this important signal molecule.

Nitric oxide is a powerful vasodilator and is locally produced by many tissues, especially endothelial tissues, under conditions of increased oxygen demand. It is then quickly eliminated (fortunately, due to its toxic effects) by reversibly binding to hemoglobin, specifically to a cysteine residue in the globin. The cysteine, thereupon contains a *S*-nitrosothiol (SNO).

The capillaries in the peripheral tissues and in the lungs are small, delicate, and constricted. One must wonder how they stay open to allow the transport of RBCs, essentially one by one. The answer lies partly with hemoglobin. In the constrained space of the capillaries, the NO is released, acts as a vasodilator in capillaries, and ensures blood pressure remains constant. Release of oxygen enables release of NO.

Chemicals That Impede Hemoglobin Functions—Poisons

Occasionally, suicide is accomplished by running an automobile in a closed space. The automobile exhaust contains carbon monoxide (CO), among other nasty chemicals that may cause cancer in the victim should the suicide attempt fail. The CO replaces oxygen in the oxygen-binding site of the heme, and the victim dies of chemical asphyxiation. Hemoglobin's binding affinity for CO is 200 times greater than its affinity for oxygen, so that small amounts of CO dramatically reduce hemoglobin's ability to transport oxygen. Concentrations as low as 0.1% CO can result in unconsciousness. You smokers inhale a surprising amount of CO. Up to 20% of your oxygen-binding sites can be occupied by CO. No wonder running is tough. Hemoglobin combined with CO forms a very bright red compound called *carboxyhemoglobin*. Should you present in a hospital emergency room with CO exposure, your treatment will be oxygen delivered at high concentrations. CO is reversibly bound and will be released when oxygen concentrations are high enough to compete with the binding sites.

Other chemicals that bind at the oxygen site on hemoglobin are cyanide (CN^-), sulfur monoxide (SO), nitrogen dioxide (NO_2), and sulfide (S_2^-), including hydrogen sulfide (H_2S). The presence of H_2S in sewer gas and other places of decaying organic matter accounts for their toxicity in confined spaces.

There have been several very well-publicized instances of water pollution that were first indicated by the appearance of *cyanosis* in young children. Cyanosis means a bluish appearance to the skin. The little ones were sensitive to the nitrates in their well water. In the body, the nitrates are metabolized to nitrous oxide which can convert hemoglobin to *methemoglobin*. Methemoglobin's iron is in the ferric (Fe^{3+}) state and cannot bind oxygen. Methemoglobin is not all bad, as it turns out. Hydrogen cyanide exerts its poisonous effect by interfering with oxidative metabolism. Hydrogen cyanide preferentially binds to methemoglobin. Victims of cyanide poisoning can be treated by chemicals that create methemoglobin, such as methylene blue, to help remove the poison from the tissues.

Disease Conditions Resulting from Abnormal Hemoglobin

There are numerous genetic anomalies that produce abnormal hemoglobin. Some of these have no or minimal effect and have only been observed through large population samples. Others have dramatic effect on the health of the individual.

The most prevalent of the latter is *sickle cell anemia*. The hemoglobin from victims of this serious genetic condition form crystalline fiber at low-oxygen tensions, such as in the capillaries. The disease results from a single point mutation that causes a valine to replace a glutamic acid on the surface of the globin molecule. The valine is nonpolar and doesn't belong on the molecular surface. In essence it forms a sticky spot and attracts other nonpolar valines on other abnormal hemoglobin surfaces, creating clumps. This condition is very prevalent in areas of high malaria incidence. It is sufficiently serious that one would expect the gene to be eliminated from the population; because its carriers, untreated, do not survive to reproduce. However, individuals who are heterozygous for this condition—in other words, have only one abnormal gene, not two—are relatively resistant to malaria. An important phase in the life cycle of the malaria pathogen occurs in the RBCs. The sickle cell conditions are demonstrably less able to support the pathogen. The heterozygous individuals have enough abnormal cells in their system to interfere with the survival of the malaria pathogen. The relative resistance to malaria of those carrying one of the anemia genes gives them a survival advantage and perpetuates the sickle cell anemia gene in the population.

Some genetic anomalies of hemoglobin act to stabilize the T state of the molecule, decreasing its affinity for oxygen. Others reduce the ability of the globin portion of the molecule to prevent oxidation of the iron, favoring the formation of methemoglobin. Some abnormal hemoglobins are unstable and precipitate out as dark conglomerates known as *Heinz bodies*. This condition results in lysis of the RBCs and consequent hemolytic anemia. If you like a good mystery, read Craig Johnson's *Death Without Company* and see if you can guess ahead of the author as to why the victim succumbed to the poison (hint—abnormal hemoglobin was involved).

In some cases, the victims experience shortness of breath and cyanosis of the peripheral tissues. Other conditions are fatal. The degree of impairment depends on the extent to which the oxygen carrying capacity of the hemoglobin is compromised. The royalty of Europe in the Middle Ages evidently carried genes for abnormal hemoglobin, perpetuated through their annoying practice of inbreeding. The impact on their health led to cyanosis (blue pallor) and reference to such individuals as "blue bloods."

Summary

There are two primary oxygen-transporting molecules, hemoglobin and myoglobin. Hemoglobin eagerly absorbs oxygen in the lung, and, just as agreeably, dispenses it to the tissues. Myoglobin resides in the tissues, primarily muscle, absorbs oxygen at a much lower pressure than hemoglobin, and releases it at a pressure lower still. Both myoglobin and hemoglobin consist of a heme secured through a pyrrole ring structure to large globulin protein subunits. Both bind oxygen by partial sharing of an electron between the iron atom and the oxygen, which is additionally stabilizing by a proximal histidine. Note that the oxygen doesn't oxidize Fe^{2+} to Fe^{3+}. The iron is protected from oxidation by the surrounding globulin. Hemoglobin molecules consist of four such subunits (tetramer) compared to one for myoglobin (monomer). Hemoglobin exists in two configurations—R (relaxed), which has high affinity for oxygen and T (tense), which has low affinity for oxygen. In the lungs, hemoglobin converts to the R configuration and takes up oxygen at about 100 torr. In the tissues, hemoglobin returns to the T configuration and releases oxygen at 20 to 40 torr. Myoglobin takes up oxygen in the 20 to 40 torr range and releases it at much lower oxygen tensions.

The behavior of hemoglobin is described by a plot of % saturation versus oxygen pressure, known as the dissociation curve. Cooperative binding gives the dissociation curve a sigmoid shape. The first oxygen to bind in a given tetramer requires high oxygen tension, but the binding of this oxygen makes binding easier for the second due to confirmation changes in the molecules. Binding of the third and fourth are easier still. Certain conditions cause the curve to shift to the right translating into less oxygen affinity and greater oxygen delivery to the tissues. This includes decrease in pH, called the Bohr effect, and binding of the molecule 2,3-BPG. Both act to stabilize the T state.

Other conditions cause a shift to the left, translating into a better facility for collecting oxygen at low pressures. Fetal hemoglobin exhibits a left shift because it is unable to bind 2,3-BPG and therefore has a high affinity for oxygen, This allows the fetus to extract oxygen at low pressures from the placenta.

The hemoglobin system optimizes both oxygen delivery and carbon dioxide elimination. CO_2 is bound physically to hemoglobin as carbamates. Also, when hemoglobin takes up oxygen in the lungs, a proton is released and reacts with HCO_3^- to release CO_2. Hemoglobin also functions in regulating vasodilation and blood pressure through selective uptake and release of NO.

Some hemoglobin poisons act by replacing oxygen at the binding site. Poisons that act in this way include CO and sulfides. Other poisons act by oxidizing the Fe^{2+} to an Fe^{3+} state. The hemoglobin containing the oxidized Fe is called methemoglobin and cannot bind oxygen. Poisons that act in this manner include nitrous oxide.

Abnormal hemoglobins usually arise from point mutations that replace a single amino acid. These may be undetectable unless they occur at the surface of the molecule. Sickle cell anemia is the most familiar and common of the hemoglobin disorders. There are others that cause methemoglobin formation. Some abnormal hemoglobins are unstable and precipitate out as Heinz bodies. This condition results in lysis of the RBCs and consequent hemolytic anemia.

PART B   Photosynthesis

Introduction to Photosynthesis

The biosphere has only one ultimate source of energy—the sun. Radiant energy can be captured only by organisms that possess photoelectric chemicals. These chemicals trap the energy of the sun and convert it to chemical energy (or potential, if you prefer) in the form of reduced carbon. All of life exists by oxidizing the carbon and capturing the enormous energy therein.

As in human-engineered photoelectric cells, radiant energy is captured when the photons falling on the sensitive chemicals elevate the energy level of electrons. You will learn how this is done as follows:

- The overall scheme by which energy is captured and carbon is reduced

- The molecule structure of chlorophyll

- The details of energy capture and production of adenosine triphosphate (ATP)

- The details of carbon reduction

- Photorespiration—a metabolic process that does more harm than good

- Poisons

The Overall Scheme by Which Energy Is Captured and Carbon Is Reduced

Electrons in the outer orbitals of an atom can be induced to jump up to the next level with the appropriate amount of energy input. However, the electrons cannot stay in intermediate stages. They must literally make a quantum leap.

Now, the electrons from any chemical can be induced to jump. Photoactive pigments, primarily *chlorophyll*, are unique because their electrons are very easily excited by photons of just the right energy—just the energy that pushes the electron to the next few energy orbital. Photosynthetic processes are designed to snatch each excited electron from the original molecule and run it through a chemical machinery that results in the production of glucose.

The energy of a photon depends on its wavelength as follows:

$$E = hv$$

where E = energy

h = Planck's constant

v = frequency (number of waves passing a given point per unit time)

Of course frequency is related to wavelength.

$$hv = \frac{hc}{\lambda}$$

where c = speed of light

λ = wavelength

Light consists of photons (little packets of energy) of a continuous spectrum of energy levels. The very short (ultraviolet) we cannot see and the very long (infrared) we cannot see. Photosynthesis uses photons of intermediate energy—those we see in the colors red and blue. The molecules absorb the red and blue photons and reflect the rest—what we see as green.

What happens when photons are captured by photoactive pigments, usually *chlorophyll*?

$$48 \text{ photons} + 18 \text{ ADP} + 18 \text{ P}_i + 12 \text{ NADP} + 12\text{H}_2\text{O}$$

$$= 18 \text{ ATP} + 12\text{NADPH} + 12\text{H}^+ + 6\text{O}_2$$

The energy from the photons is used to produce high-energy phosphate bonds, to oxidize water and to produce reducing power in the form of NADPH. As a result, oxygen is produced, for which we are all grateful. The 18 ATP translates to 2870 kJ of energy making the process a quite respectable 35% efficient. This portion of photosynthesis is called the *light reaction,* because it requires light.

Here's what the photosynthetic cell does with all this power.

$$6\text{CO}_2 + 18 \text{ ATP} + 12 \text{ NADPH} + 12\text{H}^+ = \text{C}_6\text{H}_{12}\text{O}_6 + 18 \text{ ADP} + 18 \text{ P}_i + 12 \text{ NADP}$$

Observe—in moving from carbon dioxide to glucose, carbon moves from an oxidation state of +4 to 0, thereby capturing the 2870 kJ of chemical energy. Also observe that it takes 48 photons and 18 ATPs to accomplish this. You have probably already concluded that photosynthetic cells must have mechanisms to funnel the energy from excitation of a number of electrons on a number of molecules of chlorophyll into one place. This portion of photosynthesis is called, somewhat unfortunately, the *dark reaction,* not because it necessarily occurs in the dark but because it doesn't require light.

In plants, photosynthesis occurs in the chloroplast organelle. As you may recall from Chap. 2, the chloroplast resembles the mitochondria except that the enzyme-bearing invaginations are closely packed, forming structures called thykaloids and also creating a confined space called the thykaloid lumen.

The Molecular Structure of Chlorophyll

Here, my dear readers, are the two most common forms of chlorophyll (Fig. 13-10).

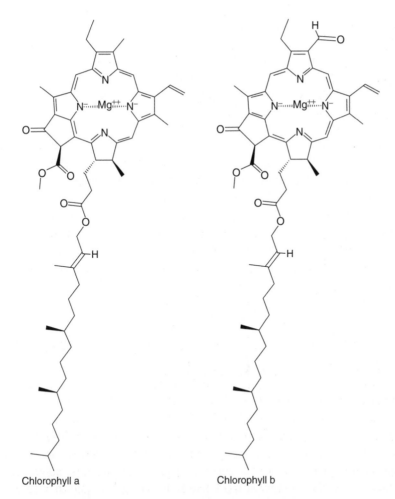

Chlorophyll a   Chlorophyll b

Figure 13-10   Chlorophyll a and chlorophyll b.

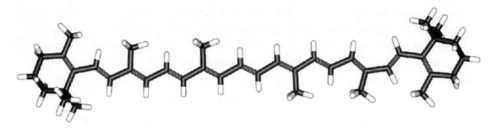

Figure 13-11   Carotene.

By golly, it looks an awful lot like hemoglobin. A metallic ion, in this case magnesium, sits in the center, surrounded by pyrrole rings. The rings are attached to globular subunits. There is also a long aliphatic chain that serves to keep the chlorophyll where it belongs—inside the stoma (see Chap. 2). The whole thing is designed to absorb photons of the red/blue frequency.

Note that there are two major types of chlorophyll, a and b. They have slightly different absorption spectrums, and activation of both is necessary for one turn of the mechanistic wheel to produce 1 molecule of glucose.

There are some other light-absorbing compounds, known as *pigments*, that are additional to the chlorophyll, hence called *ancillary pigments*. These include the famous carotenoids, which are orange like a carrot. These pigments mop up photons at different energies than chlorophyll and give the whole process a little boost. In the fall, when the chlorophyll is drained from the leaves, you are provided a glimpse at the ancillary pigments. See Fig. 13-11.

The ancillary pigments actually appear to be an evolutionary improvement upon chlorophyll. The peak of solar radiation is at the green/yellow wavelength interface. A pigment that reflects green light, like chlorophyll, is missing the spectrum with the greatest energy content. Plants that are not green are more sophisticated, I'll have you know.

The Details of Energy Capture and Production of ATP

Sunlight is a diffuse energy form. Photo-captivating biosystems are designed to concentrate this energy. Numerous molecules of chlorophyll and the ancillary pigments, collectively called *antenna pigments*, are clustered together in areas called *light harvesting centers (LHC)*. The LHCs are dense and therefore likely to capture whatever light falls upon them. The energy contained in the excited electrons is funneled into pairs of chlorophyll molecules that have the mission of initiating photosynthesis. These chlorophyll pairs are called *reaction centers (RCs)*.

In the reaction centers of plants and cyanobacteria, light-driven oxidation of H_2O produces NADPH and ATP. This occurs through the action of three protein complexes that span the thykaloid membrane—*photosystem I (PSI), photosystem II (PSII)* and the *cytochrome bf complex*. The oxygen-producing part of photosynthesis occurs through the actions of two photosynthetic RCs acting in series.

We will start with PSII. Electrons are activated in special chlorophyll molecules that absorb light of 680 nm wavelength, thereby named P680. With the activated electrons, they are denoted P680*. Usually, excitation of an electron results in one of two things— the electron falls back to its normal energy state and emits light (fluorescence) or the electron falls back to its normal energy state and emits heat. Before either of these useless events occurs in the reaction center, the activated electron is snatched from the original molecule. These electrons cascade down a series of molecules with the ultimate consequence of oxidizing water, generating oxygen and protons. The cascade results in the pumping of protons into the thykaloid lumen which creates a pH gradient between the lumen and the chloroplast stroma. Here is where oxygen is evolved. This activity is schematically depicted in Fig. 13-12.

P680* will pass either one or two electrons to a ubiquinone analog called *plastoquinone*. The plastoquinone picks up two protons from the stoma along with the two electrons. The electrons are passed to a *plastocyanin* molecule and the two

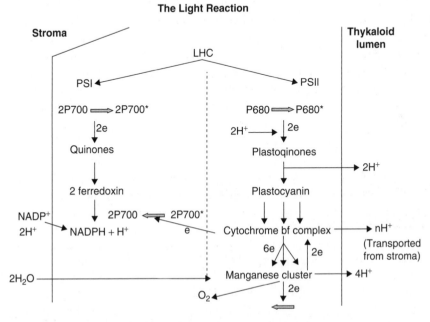

Figure 13-12 The light reaction.

protons are pumped into the thykaloid lumen. The stroma becomes more basic and the lumen more acid.

Now the electrons move to the cytochrome bf complex. Cytochrome bf contains a heme that can manipulate electrons by alternating the valence states of its Fe. The excited electrons, together with electrons from oxidation of water, are used to drive movement of additional protons from the stroma to the lumen.

Now the electrons move to the *manganese cluster,* also called the *oxygen evolving center* (OEC) because this manganese-containing cluster oxidizes water and evolves oxygen.

$$2H_2O + \text{manganese center} = O_2 + 4H^+ + \text{reduced form of manganese center}$$

Note that more protons are generated that move into the thykaloid lumen.

Finally, a system called the "Z" system restores electrons to the P680*. The PSII has done its job—creating a proton gradient across the thykaloid membrane.

Now let's look at PSI. The job of PSI is to create reducing potential. Here's how it is done. The RC is composed of chlorophyll called P700 because of its preference for light of wavelength 700 nm. The activated P700 is deemed P700*. The electrons are transferred to a series of quinones and then to ferredoxin. The ferredoxin uses its reducing power in the form of reduced Fe to convert $NADP^+$ to NADPH (nicotinamide adenine dinucleotide and its reduced form). When the P700* picks up an electron from the cytochrome bf complex, the job is done. This is all summarized in Fig. 13-12.

At this point, no energy has been generated. Let us return to the proton gradient across the thykaloid membrane. Chloroplasts and mitochondria capture energy the same way. Could they have a common ancestor? They both couple the dissipation of a proton gradient to the enzymatic synthesis of ATP. Unlike the membrane of the mitochondria, the thykaloid membrane is freely permeable to molecules like Cl^-, so the gradient is purely pH, not electrochemical. When the protons in the lumen move down the gradient, energy is released and harnessed into ATP molecules by the enzyme *ATP synthase*. The process is called photophosphorylation. Again, the 18 molecules of ATP produced translate to 2870 kJ of energy.

The process is somewhat different for prokaryotic compared to eukaryotic cells, because the prokaryotic cells do not have the convenience of organelles. However, they do form *chromatophores* from invaginated structures of the plasma membrane. The proton gradient is established across the invaginated plasma membrane into the periplasm.

The reaction centers of photosynthetic prokaryotes contain slightly different chlorophylls than eukaryotes. The reduction equivalents of prokaryotes are not derived necessarily from water but also from H_2S, S, S_2O, and H_2. A typical reaction is:

$$CO_2 + 2HS = CH_2O + 2S + H_2O$$

Organisms that perform this reaction are the oldest on earth. By the metric of longevity, they are the most successful—except that they have been marginalized into inhospitable environments by the evolution of oxygen and the development of an oxidizing atmosphere. By the metric of diversity and adaptability, they are the least successful.

The Details of Carbon Reduction

Carbon reduction, the dark reaction, is also called the *Calvin cycle*. The enzyme that accomplishes the reduction of carbon is the most important protein on earth. Its name is *ribulose bisphosphate carboxylase/oxygenase* or, fondly, *RuBisCO*. RuBisCO catalyzes the addition of a carbon from carbon dioxide to phosphorylated ribulose, which breaks into 2 molecules of 3-phosphoglycerate.

Remember that you are trying to get a 6-carbon sugar. Here is how it is done. The process is depicted in Figs. 13-13 and 13-14.

In summary, a carbon atom is added to a phosphorylated, five-carbon carbohydrate, resulting in a 6-carbon carbohydrate. This 6-carbon carbohydrate breaks into two 3-carbon molecules. The next step requires energy in the form of ATP whereupon the 3-carbon carbohydrates pick up an additional phosphate group each. This step— done once for each of the two carbon entities—converts 3-phosphoglycerate to 1,3-bisphosphoglycerate. The next step—done once for each of the two carbon

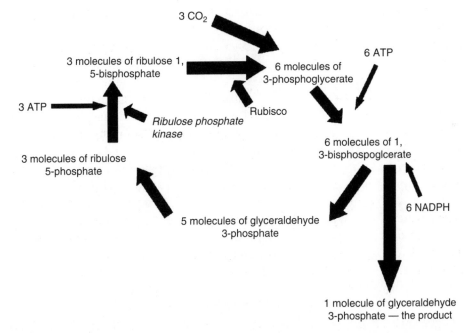

Figure 13-13 The Calvin cycle.

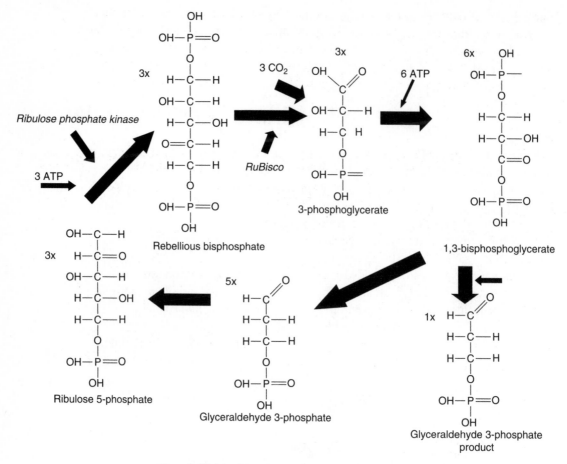

Figure 13-14 The Calvin cycle—another view.

entities—requires use of reducing power in the form of NADPH, which reduces one of the phosphate groups to hydrogen, creating glyceraldehyde-3-phosphate. Because each turn of the cycle fixes 1 molecule of carbon dioxide, the production of 1 molecule of glucose, formed from newly reduced carbon, requires six turns of the Calvin cycle.

So, at this point, you have consumed all of the reducing power and three-fourths of the energy generated from the light reaction. The glyceraldehyde 3-phosphate is converted into fructose-6-phosphate for entry into gluconeogenesis.

Six cycles × 1 ATP per phosphorylation of 3-phosphoglycerate
× 2 molecules of phosphoglycerate = 12 ATP

The remaining 6 molecules of ATP are used in the regeneration of ribulose biphosphate.

Ribulose phosphate kinase is required for the rejuvenation of ribulose bisphosphate. This enzyme will not function unless a cystine disulfide within the molecule is reduced to cysteine. Therefore, if excess reducing power is not available, indicating that NADPH is not being transferred from the light reaction, the dark reaction will not go forward. Ribulose phosphate kinase provides a clever coupling of the light and dark reactions.

Photorespiration—A Metabolic Process That Does More Harm Than Good

Here's a mystery—oxygen competes with carbon dioxide at the active site of RuBisCO. If oxygen initiates the metabolic process, the ribulose biphosphate is oxidized to phosphoglycolate and no energy is captured in the form of reduced carbon. The process of photorespiration is shown in Fig. 13-15.

It is probable that the dark reaction of photosynthesis was developed in the days before the earth had an oxidizing atmosphere. The accumulation of oxygen has set the process back. The interference of the dark reaction by the presence of oxygen takes a significant toll on the efficiency of photosynthesis. One class of plants has evolved a system to minimize the occurrence of photorespiration. These plants are called C4 plants because their dark reaction captures the carbon as a 4-carbon compound. The length of the initial molecule distinguishes this class of plants from the majority of photosynthetic organisms on the plant. Most plants add carbon to form a 3-carbon entity (3-phosphoglycerate) and are therefore called C3 plants. The dark reaction in C4 plants is shown in Figs. 13-16 and 13-17.

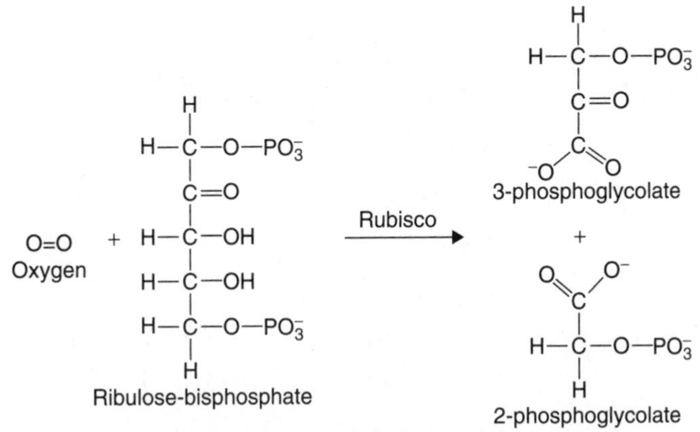

Figure 13-15 Photorespiration.

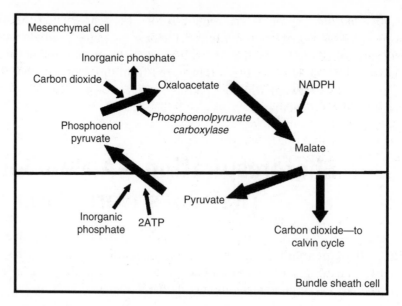

Figure 13-16   C4 dark reaction.

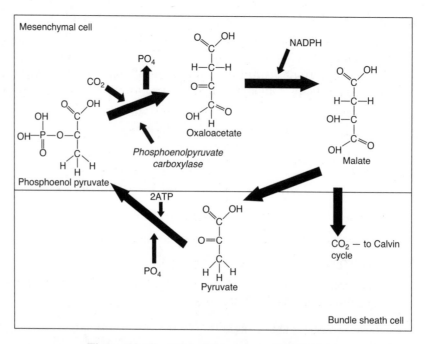

Figure 13-17   C4 dark reaction—another view.

The C4 plants first capture the CO_2 on the surface of the leaf in *mesophyll* cells using a system different than RuBisCO. The C4 plants first capture the new carbon as oxaloacetate using the enzyme *phosphoenolpyruvate carboxylase*—an enzyme that has a high affinity for CO_2 and a low affinity for O_2. The oxaloacetate is transported deeper into the leaf where oxygen tension is low. It moves into *bundle sheath cells*, where the CO_2 is liberated and given to RuBisCO. This process takes 2 ATP. It seems that the C4 plants should be the more numerous, but they are not. C4s, mostly grasses, constitute only about 5% of the plant population. I suspect that, for all its faults, direct use of RuBisCO has its advantages. It is of note that attempts to improve RuBisCO function, pointedly to decrease affinity for oxygen by bioengineering, have not been successful.

Summary

Photosynthesis is the only process on earth which makes energy available to biosystems. The energy is provided by solar radiation. Photosynthesis converts one form of energy (radiant) into another (chemical). The radiant energy is captured by photoactive pigments, the most numerous of which are types of chlorophyll. The chlorophylls absorb radiation in the blue/red range, reflecting the light we see as green. There are other light-gathering molecules, called pigments because of their colorful nature, which absorb light of different frequencies than chlorophyll. These are ancillary pigments and their existence is evident in the fall, after the chlorophyll molecules have disappeared.

The light energy is captured within structures called light harvesting centers (LHC), full of chlorophylls and other pigments. The energy in the form of excited electrons is channeled to reaction centers where photosynthesis occurs. Photosynthesis occurs through two distinct systems, one that produces reducing power in the form of NADPH and one that produces ATPs, described as PSI and PSII respectively. In eukaryotes, the production of ATP requires establishment of a proton gradient across the thykaloid membrane of the chloroplast.

Quiz

1. The sigmoid shape of the oxyhemoglobin dissociation curve is due to

   (a) decreased pH at increasing oxygen tension.

   (b) the moping up of released oxygen by myoglobin.

   (c) cooperative effect of binding one oxygen molecule on the binding of subsequent oxygen molecules.

   (d) drop of oxygen pressure as oxygen is taken up.

   (e) None of the above.

2. Myoglobin differs from hemoglobin in

    (a) the shape of the heme.

    (b) the shape of the protein surrounding the heme.

    (c) the oxidation state of the iron.

    (d) the allosteric effect of oxygen binding.

    (e) All of the above.

3. The effect of lower pH on oxygen binding

    (a) increases the ability of fetal hemoglobin to extract oxygen.

    (b) decreases delivery of oxygen to the peripheral tissues.

    (c) stablizes the R state.

    (d) is known as the Bohr effect.

4. Hb enables transport of carbon dioxide to the lungs by

    (a) physically binding the carbon dioxide.

    (b) catalyzing the formation of HCO_3^-.

    (c) lowering the pH.

    (d) releasing NO.

5. Ferric (Fe^{3+}) state of hemoglobin iron means

    (a) increased ability to bind oxygen.

    (b) formation of methemoglobin.

    (c) increased delivery of oxygen to the peripheral tissues.

    (d) All of the above.

6. The energy for photosynthesis

    (a) is provided by photons in the green spectrum of light.

    (b) is funnelled from the excitation of electrons on numerous pigments into a single molecule of RuBisCO.

    (c) is gathered by antenna pigments and funnelled into RCs.

    (d) is used to activate an electron on CO_2, allowing its conversion to CH_2O.

    (e) (c) and (d) are correct.

7. Photosynthesis differs from oxidative metabolisms in that

    (a) a MgII ion is oxidized instead of FeII.

    (b) NADP+ is the ultimate electron acceptor rather than oxygen.

    (c) it doesn't depend on the establishment of a proton gradient.

(d) it is not a redox reaction.

(e) None of the above.

(f) All of the above.

8. Photorespiration is the process by which

   (a) oxygen competes with carbon dioxide for RuBisCO.

   (b) energy from the sun is captured to ultimately fuel the production
        of ATP.

   (c) plants metabolize glucose to provide energy for growth.

   (d) plants evolve oxygen from reduction of CO_2.

9. Production of ATP

   (a) occurs in the stroma of the chloroplast.

   (b) is fueled by disassembling of a proton gradient across the thykaloid
        membrane.

   (c) requires ATP synthase.

   (d) All of the above.

10. The electron from P680* is restored by

   (a) oxidation of the manganese cluster.

   (b) receipt of another excited electron from the LHC.

   (c) PSI.

   (d) None of the above.

CHAPTER 14

Bioinformation, Analytical Techniques, and Bioinformatics

Introduction

The ability to actually see the genetic structure of various organisms and the understanding of how this structure translates into the function of the cell has truly revolutionized our view of life. A new age in science has dawned with the arrival of a brand new area of science called *bioinformatics*. Bioinformatics is the analyses of

data related to DNA sequences, called *genomes*, and also of related information on the structure and function of proteins.

The massive amounts of data that forms the basis of bioinformatics can only be collected because of the availability of computerized analytical tools. Can you image trying to find patterns among 8000 sequenced genomes without the use of searchable electronic databases? One of the really smart things done early in the work on DNA sequences was the implementation of international databases to house both genetic information and information on protein structure and function.

In this chapter, we will learn

- The basic science behind the production of proteins by the genetic code
- Methods used to study protein structure and function
- Methods used to determine the base code sequence on DNA
- Analytical techniques in deriving relationships between genomes

Production of Proteins

As you know, DNA consists of strands of nucleic acids, comprising the nucleotides adenosine (A), thymine (T), guanine (G), and cytosine (C). The sequence of these compounds constitutes a code. Three nucleic acids in a row are a code for a specific amino acid. For example, the sequence ATA codes for tyrosine. The sequence of three nucleic acids that code for an amino acid is called a *codon*. A string of 300 codons would instruct the cell to build a protein consisting of 300 amino acids. Table 14-1 gives the genetic code that is used almost universally throughout all life forms (some mitochondrial and some archaeal codes being the exceptions).

You may be surprised that the code for all of life can be encoded within only four different signals. Early researchers were surprised also. Until the middle of the twentieth century, most scientists assumed life was encoded in the proteins. After all, proteins have 20 different signals. However, within the 300 codons that call out amino acids for an average protein are 900 positions for organic bases. Each position has four possibilities. Because order matters, the number of different combination of 900 positions with four options is 10^{40}. That's enough different proteins to do us for now.

You will note some peculiarities about this code. For one thing, it is redundant. Most amino acids have at least two codons. Usually, the codons differ only in the third position. Therefore, if the DNA experiences a mutation at the spot of the third base, the cell is not affected because the same amino acid is indicated. Also notice that the code contains start and stop instructions so that the process begins and ends at the right points to cause the formation of a protein with the right sequence of amino acids.

Table 14-1   DNA Codes for Amino Acids

Amino Acid	mRNA Code
Alanine	GCU, GCC, GCA, GCG
Arginine	AGA, AGG, CGU, CGC, CGA, CGG
Asparagine	AAU, AAC
Aspartic acid	GAU, GAC
Cysteine	UGU, UGC
Glutamine	CAA, CAG
Glutamic acid	GAA, GAG
Glycine	GGU, GGC, GGA, GGG
Histidine	CAU, CAC
Isoleucine	AUU, AUC, AUA
Leucine	CUU, CUC, CUA, CUG, UUA, UUG
Lysine	AAA, AAG
Methionine	AUG
Phenylalanine	UUU, UUC
Proline	CCU, CCC, CCA, CCG
Serine	UCU, UCC, UCA, UCG
Threonine	ACU, ACC, ACA, ACG
Tryptophan	UGG
Tyrosine	UAU, UAC
Valine	GUU, GUC, GUA, GUG
Stop	UAA, UAG, UGA
Start	AUG

The code is read by RNA. By read, we mean that a single strand of RNA is formed that is complimentary to the code for a protein on a section of the DNA. The formation of the complementary RNA strand is achieved by the enzyme RNA *polymerase*. The RNA polymerase attaches to a specific configuration of the DNA. This configuration requires the presence of activation factors, such as some hormones, and results in the readout of the genetic code for a specific protein. A *promoter* is a short sequence of DNA bases that is recognized as a start signal by RNA polymerase. The RNA polymerase unwinds the portion of the DNA containing the code of interest and assembles the complementary RNA molecule.

RNA, like DNA, respects complimentary base pairing. However, RNA contains uracil instead of thymine. So the pairs are A-U and G-C. The code of ATA in the example above would result in the formation of an RNA molecule that contained the sequence UAU. The type of RNA that picks up the code from the DNA is called *messenger RNA (mRNA)*. The process whereby the mRNA is formed is called *transcription*. See Fig. 14-1.

Once formed, the mRNA carrying the genetic code for a specific protein leaves the nucleus. The mRNA attaches to a small subunit of a ribosome. The ribosomes are the assembly points for proteins. After the attachment of the mRNA to the small subunit of the ribosome, the large subunit then attaches; and protein production begins. The ribosome behaves like a rachet, moving down the mRNA, codon by codon, and assembling the protein. The process of assembling the protein from the mRNA is called *translation*.

The amino acids are collected by another type of RNA called transfer RNA (tRNA). See Fig. 14-2. Transfer RNA has been called the "Rosetta stone of life" because these

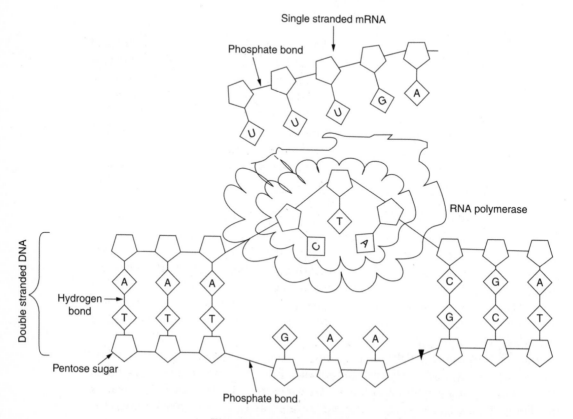

Figure 14-1 Transcription.

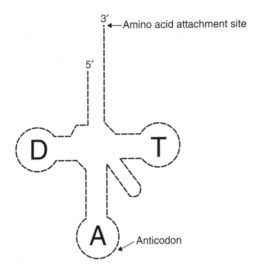

Figure 14-2   Transfer RNA.

molecules interpret the genetic code brought to the ribosome by the mRNA and provide the designated amino acid. The tRNAs are specialists; each tRNA is specific to a given amino acid. The tRNA contains a segment that is complementary to the code on the mRNA. Consider the example of tyrosine. The codon for tyrosine on DNA is ATA. The messenger RNA carries the complementary code of UAU. The tRNA for tyrosine not only binds tyrosine but also contains a segment of AUA. The AUA is called an *anticodon* and binds to the UAU on the mRNA. Tyrosine is added to the protein at the appropriate spot according to the code on the mRNA. The process whereby the code on the mRNA is translated to produce a protein is cleverly called translation. See Fig. 14-3.

From the viewpoint of the researcher, the process of translation is unnecessarily complex. From the viewpoint of nature, perhaps, it is just right complex. The code for any given protein is interspersed with DNA that does not belong. The sections of the code that are specific for amino acids are called *exons* (because they are expressed) and the sections of the DNA that are excess are called *introns*. The RNA polymerase faithfully duplicates the whole thing. After transcription, the mRNA is edited. The unnecessary sections are cut out by catalytic RNA molecules called *splicesomes*.

One of the mysteries placed in nature to entertain biochemists is the fact that there are only about 30,000 genes within the human genome. The reason this is mysterious is because there are over 100,000 different proteins. It may be that a specific mRNA can code for different proteins depending on how it is edited.

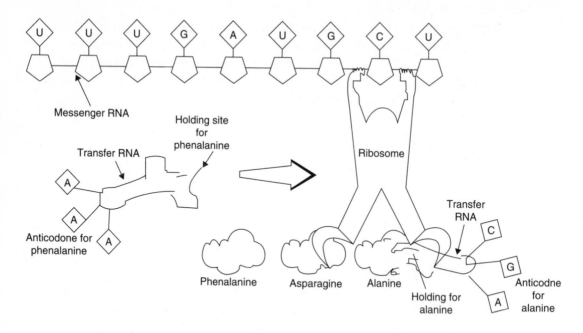

Figure 14-3 Translation.

Methods Used to Study Protein Structure and Function

Among the disciplines related to bioinformatics is *proteomics*, the study of the protein complement of various cells and organisms. The entire complement of proteins within an organism is called the *proteome*. The goal of proteomics is to characterize the *phenome*. The phenome is the sum total of all of the physical characteristics of a given organism. Proteomics seeks to understand the function of various proteins found within the cell and also the relationships between proteins within a given organisms and among organisms.

There are a number of diagnostic techniques used to characterize proteins, including chromatography, gel electrophoresis, mass spectrophotometry, x-ray crystallography, and column chromatography. Each of these is briefly described below to give you, dear reader, a high level of understanding of the information to be gained.

Someday you, as a gifted biochemist, will be in a position to possibly obtain a patent because you have isolated and characterized a new protein. Yes, one can obtain patents on biomolecules. One school of thought believes that allowing biochemists to patent biomolecules is similar to allowing physicists to patent gravity. However, the knowledge of the structure and function of a protein can translate into a lot of money for someone—especially if the protein is important in human nutrition, disease, or in efficiency of production of a crop. The regulatory community has taken the position that such knowledge must be patentable to stimulated invest in this type of research and motivate researchers, investors, and corporations.

SEPARATING PROTEINS IN A MIXED SAMPLE

Consider that any sample collected from a living organism will contain a bewildering assortment of molecules. How do you isolate a protein of interest from this mess? There are two commonly used methods to separate molecules on the basis of size and/or chemical characteristics. These are

1. Electrophoresis
2. Chromatography

Electrophoresis uses an electric field to isolate molecules of different sizes, usually by pulling a protein mixture through a stationary media, or gel. Proteins to be separated solely on the basis of size must be denatured. Remember that a denatured protein relaxes into a random coil, losing all but its primary structure. Agents such as 2-mercaptoethanol or dithioreitol are used to reduce the disulfide bridges holding proteins in their tertiary configuration. Then the proteins are further denatured with sodium dodecyl sulfate (SDS), an anionic detergent that wraps around the polypeptide chain and confers upon it a negative charge. Because all of the proteins wrapped in SDS exhibit the same ionization, they migrate through the electric field with the same strength. However, the bigger molecules are impeded by the gel and move more slowly. As a result, the different molecules segregate within the gel, based on size. A linear inverse relationship exists between the logarithm of the molecular weight of an SDS-denatured polypeptide and the ratio of the distance migrated by the molecule to that migrated by a standard marker dye-front. The researcher uses a standard curve of distance migrated versus $\log_{10}$ of the molecular weight for known samples to determine the molecular weight of various proteins within the sample after measuring distance migrated on the same gel.

To see where the molecules are within the gel, they must be labeled somehow. Commonly, the proteins are tagged with a radioactive label, such as tritium (^{3}H) or carbon-14 (^{14}C). A photosensitive film is laid upon the gel and the radioactive proteins appear as areas of exposed film. Alternatively, the proteins are tagged with a fluorescent material that can be induced to fluoresce in the gel, revealing the position of the molecules. In some systems, the gel electrophoresis is followed by blotting and treatment with labeled antibodies specific to the protein of interest. See Fig. 14-4.

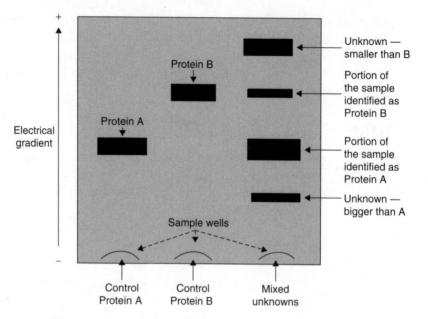

Figure 14-4   Gel electrophoresis of a mixed protein sample.

If the sample is not exposed to SDS, then the relative movement of the material is also affected by the degree of ionization of the polypeptide chain. A specialized version of non-SDS electrophoresis is called *isoelectric focusing*. The isoelectric point of a given ion is the pH at which its net charge is zero. Because the net charge is zero, mobility in an electric field is zero. An uncharged molecule will not move any further in an electrophoresis system. If the electrophoresis gel presents a pH gradient, molecules will stop movement at the pH which represents their isoelectric point. See Fig. 14-5. To study individual molecules, the gel is literally cut into areas containing the molecules of interest and submitted to further analysis.

Proteins can also be separated on the basis of size by *chromatography*. The chromatography sample is called the *liquid phase* and the material it moves through is called the *stationary phase*. In column chromatography, the sample is introduced onto the top of a column that is packed with beads of polystyrene or other materials. Typically, the beads are perforated with small holes. The column is then eluted with an appropriate eluent. Unlike electrophoresis, the smaller molecules wash through the matrix more slowly because they tend to hang up in holes in the beads and in the interstitial spaces between the column beads.

The matrix that the sample moves through doesn't have to be beads and it doesn't have to be held in a column. In thin-layer chromatography, a liquid sample is introduced into a matrix that has been applied as a thin layer, usually between two glass plates. The liquid moves through the matrix by capillary action. In some systems,

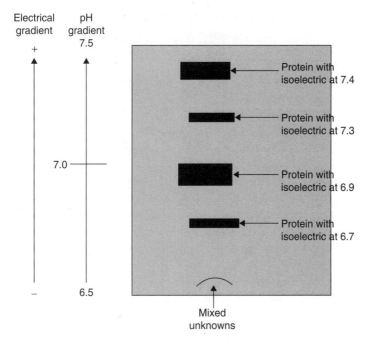

Figure 14-5   Isoelectric focusing.

the sample is converted to a gas prior to passage through a liquid substrate. Some systems are even based on a paper stationary phase.

Chromatography can also be used to separate a sample on the basis of chemical properties. For example, in *ion exchange chromatography*, the stationary stage carries a charge, either negative or positive. Charged molecules in the sample pass through the column until they encounter a binding site in the column. The column is then eluted with a substance of opposite charge or with eluents of different ionic strength. The tendency of a molecule to move through the column will vary with the ionic strength of its attraction to the stationary phase.

A series of samples is taken sequentially as eluent moves through a chromatography system. The time at which an individual molecule emerges from the chromatography column is indicative of its relative size, and, sometimes, its chemical properties. As with gel electrophoresis, the molecules have to be labeled with something, so that you can see or detect them. See Fig. 14-6. These samples, hopefully of individual proteins, are then available for further analysis.

Other ways to separate molecules within a sample include ultracentrifugation, wherein the sample undergoes centrifugation in a system that causes the sample to stratify depending on the size of the molecules. Another separation method is called salting out. Salting out is based on the fact that different proteins will form precipitates at different salt concentrations, dependent partly on their ionic charge.

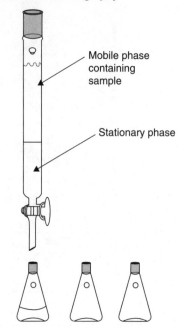

Figure 14-6 Chromatography.

DETERMINING THE AMINO ACID CONTENT OF AN ISOLATED PROTEIN

A favorite technique for determining the amino acid content of a protein, once it has been isolated, is *mass spectrometry*. The use of a mass spectrometry depends on the conversion of the molecule to ions, a step that can be quite tricky for a biomolecule. Once ionized, the polypeptides are fragmented. Ionization is commonly achieved by bombarding the sample by a high-energy beam of electrons. The ions are then accelerated within an electric field, which is bent by a perpendicular external field. The position of the ions is detected electronically. The speed and direction of movement depends on the charge (z) and the mass (m) of the ion. The data is presented as a ratio m/z. The signature m/z of each amino acid has been previously established. Mass spectrometry will provide you with the type and amount of amino acids in the polypeptide. See Fig. 14-7 for an example. The relative size of each peak is proportional to the relative amount of the identified amino acid.

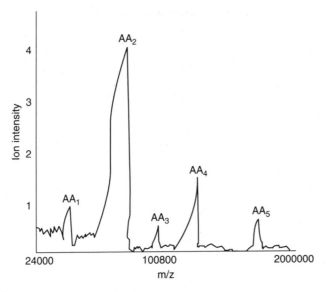

Figure 14-7   Example results from mass spectrometric analysis of a peptide.[*]
[*]The amino acids occur in the sample protein with the following relative frequencies:
$$AA_2 > AA_4 > AA_1 > AA_5 > AA_3$$

SEQUENCING AMINO ACIDS

In order to characterize a protein, you need to know order of appearance of the amino acids. *Tandem mass spectrophotometry* applies multiple steps of mass spectrophotometry separated by some form of fragmentation. A computerized algorithm can take the results from the multiple runs and construct a probable amino acid sequence for the protein in question.

DETERMINING PROTEIN STRUCTURE

As you know, the amino acid sequence within a protein is only part of the story. The function of a protein depends on its shape. The three-dimensional structure of proteins is determined by the technique of *x-ray crystallography*. This technology reveals the arrangement of atoms within a crystal by the pattern generated when a beam of x-rays is passed through the crystal. The x-rays are scattered by the electrons contained in the molecule. The pattern exhibited by the scattered x-rays allows the researcher to deduce the density of electrons within the atomic structure and also the nature of their chemical bonds. The molecule has to be converted to a crystal for this to work, a mean trick for biomolecules.

Among the first structures to be elucidated by x-ray crystallography were cholesterol (1937), vitamin B_{12} (1945), and penicillin (1954). The chemist who analyzed all three, Dorothy Crowfoot Hodgkin, was awarded the Nobel Prize in Chemistry in 1964. Just to show you how hard this is, Dr. Hodgkin spent the next 30 years solving the structure of insulin. Nonetheless, over 39,000 x-ray crystal structures of proteins, nucleic acids, and other biological molecules have been determined.

A newer technique of structure analysis is *nuclear magnetic resonance spectroscopy* (usually abbreviated NMR). This technique is based on the fact that within a magnetic field, NMR active nuclei (such as ^1H or ^{13}C) resonate at a frequency characteristic of the isotope. NMR studies magnetic nuclei by aligning them with a magnetic field and then interrupting this alignment with an alternating magnetic field. The resulting response yields characteristic high, resolution spectra that are exploited to reveal molecule structure.

DETERMINING PROTEIN FUNCTION

The cell proteome is enormous. We understand the function of only a fraction of the proteins that are produced. Once you have isolated a protein and determined some of its physical characteristics, then what? How can you determine what the protein does? The tools of the science of bioinformatics are there to help you. Massive databases are available containing all known information (assuming no one is holding out on us) about proteins, their structures, and their functions. Using powerful search engines, you can find proteins with similar structures. Remember that many proteins contain common structural themes, called domains. You would be especially interested in proteins that had domains similar to those found in your mystery protein. The functions performed by proteins with structures analogous to yours might give you some idea of what your specific protein does.

One way of deducing the function of a protein is to eliminate the protein from the cell proteome. *Antisense* technology will allow you to eliminate your protein from the cell where you found it. Refer back to the genetic code that calls out specific amino acids. Using this information, and the sequence of amino acids on your protein sequence, you can determine the complementary sequence on the mRNA that cell uses to produce your protein. Remember that RNA molecules are single stranded. An antisense strand of RNA can be constructed that contains the organic base sequence complementary to the target mRNA. This molecule will bind to the single-stranded mRNA and prevent it from producing your protein. See Fig. 14-8. You can then observe the physiological impact of losing the protein.

Alternatively, you can analyze for the protein under different conditions of physiological stress. If this protein appears only when the cell is experiencing stress of a specific sort, you can conclude that the protein helps the cell respond to this stress; if you suspect that this particular protein is helpful under specific environmental

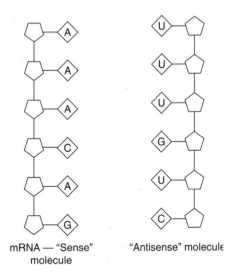

mRNA — "Sense"   "Antisense" molecule
molecule

Figure 14-8   Depiction of an mRNA molecule and its antisense molecule.

conditions, you can introduce the gene for the protein using bioengineering technology into cells under the suspect conditions. Needless to say, this is a very resource-intensive endeavor. You need an appropriate vector to introduce the gene into the cell and you need an appropriate cell that will express the gene and utilize the protein.

One of the most widely used ways of determining the function of proteins is *DNA microarrays* or *gene chips*. Microarrays identify which proteins are being produced under different environmental conditions. They are literally chips with embedded DNA fragments. The DNA fragments are arranged in an array in a solid matrix. Microarrays created from the DNA material of all types of organisms and of human cells are available commercially. These microarrays are fundamental in research to determine which genes are expressed within given cells. They are very important in determining the genetic basis for the difference between normal and abnormal cell function. See Fig. 14-9.

To understand how microarrays are used, consider the researcher who is studying the difference in cell function between pancreatic cells of normal individuals and individuals suffering from diabetes. They will harvest mRNA from both types of cells. The mRNA is used to produce a gene chip and therefore can tell the researcher which proteins are being produced. Converting the mRNA to DNA requires the use of an enzyme called *reverse transcriptase*. Reverse transcriptase is found in retroviruses, Retroviruses enter the cell as RNA. The reverse transcriptase causes the host cell to produce the complementary DNA from the virus RNA and then insert the viral DNA into the host DNA. Reverse Transcriptase can be used in a laboratory environment to produce DNA from any type of mRNA. The DNA produced from

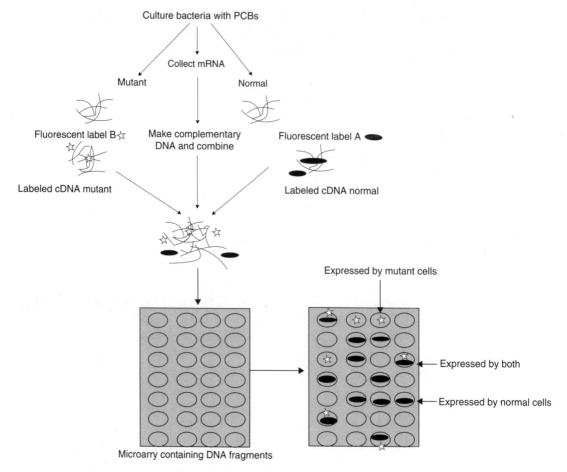

Figure 14-9 Using microarray technology.

mRNA is called *complementary DNA* or cDNA. The researcher in this example will use the mRNA harvested from a cell of interest to produce cDNA.

The cDNA from the normal cell will be labeled with one fluorescent label, say green for the sake of illustration. The cDNA from the abnormal cell will be labeled with a different fluorescent label, say blue. The researcher will obtain a DNA chip with denatured DNA fragments from the genome of pancreatic cells. The DNA fragments are arranged in an array on a microscope slide. The location and identity of each fragment is known. For example, in the location two places down from the top and three places up from the left, the researcher might know that a gene fragment with the sequence A-T-C-G-C is present. The microarray is incubated with the cDNA made from mRNA from both the normal and pancreatic cells. The cDNA will be fixed on the matrix by binding with the denatured DNA fragments that contain the matching (or mirrored) base pairs. By scanning the array with first a green and then a blue laser, the researcher will know which genes are expressed by both cells, which genes are expressed by only the normal cell and which genes are expressed by only the abnormal cell. The microarray is like a switchboard. The operator can tell at a glance which circuits are turned on.

Using DNA microarrays, you can determine conditions that involve production of your protein and perhaps deduce the function of the protein.

You must certainly suspect that our simple model of a single protein implementing a function within the cell physiology is too simple. Some proteins are needed to induce the formation of other proteins. The first protein acts as an activation factors for readout of the second protein. Indeed, many functions involve coordinating a series of enzymes. One of the techniques to study the interactions between proteins is called "bait and prey." Genes for the proteins under investigation are introduced into host cells, usually yeast, on different vectors. Cells are grown on media that requires expression of a combination of the genes on different vectors. Only cells with the correct combination of proteins will be able to survive.

Determining Base Code Sequences on DNA

The *genome* is the entire genetic complement of an organism, that is, all the organic bases contained within the DNA. It is important to remember that the genome contains much more than genes; in fact, for advanced organisms, the majority of the genome is *not genes*. Late in the twentieth century, the genomes for several small organisms were determined, including fruit fly and *E. coli*. With the dawn of the twenty-first century, the DNA sequence for the entire human genome was revealed. Genomes for hundreds of additional species have subsequently become available. The huge amount of data being generated will entertain biochemists and geneticists for generations to come. But how was this miraculous accomplishment done? There are several key technologies involved in sequencing and otherwise studying DNA.

- Polymerase chain reaction
- Southern blot
- Nuclear probes
- Sanger technique
- DNA synthesizers

POLYMERASE CHAIN REACTION

The *polymerase chain reaction (PCR)* makes many copies of a small fragment of DNA. This technology has been instrumental in the successful completion of the human genome project as well as other applications, such as DNA analysis to identify perpetrators of crimes, or to indemnify suspects of crimes. PCR essentially duplicates the process used in the cell to make copies of DNA.

The target genetic material is heated to 90 to 96°C which denatures the DNA and causes it to unwind and separate. In the second step, short segments of complementary bases, called primers, are attached to the end of the now single-stranded DNA. In the third step, an enzyme called a *polymerase* reads each template strand and quickly matches it with complementary nucleotides. The polymerase will make two new DNA double strands out of the original one. With this technology, a small segment of DNA can be used to generate enumerable copies. See Fig. 14-10.

PCR is conducted at temperatures that denature most enzymes, including most polymerases. A major component in the successful implementation of PCR is availability of a form of polymerase that is tolerant of high temperatures. This polymerase, called *Taq*, was isolated from *Thermus aquaticus* organisms that dwell in hot springs. Taq is functional in the rapidly changing thermal environment of the automated PCR process.

SOUTHERN BLOT

The Southern blot technique was named after its developer, Edward Southern, and is used to analyze DNA fragments. To apply the Southern blot technique, the DNA segment is digested into small fragments. Cutting DNA into pieces is best done using *restriction enzymes*. To DNA researchers, restriction enzymes are the best things since sliced bread. Restriction enzymes are used to insert genes into genetic

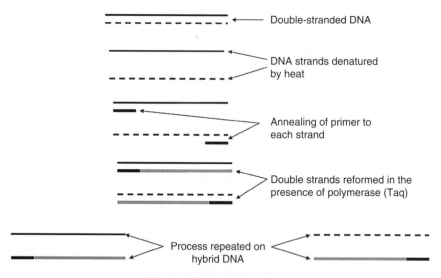

Figure 14-10   PCR techniques.

material. These nifty little enzymes cut DNA at predicable base sequences. Each type of enzyme has its own favorite base sequence. Restriction enzymes were first isolated from bacteria where they function to disable invading viruses.

The digested DNA fragments are submitted to gel electrophoresis. Each fragment is placed in a well along the edge of the gel. As you recall, the speed through the gel depends on the size of the molecule, because the gel is actually a matrix of interlocking fibers. Smaller compounds find it easier to navigate through the matrix than do larger compounds. To visualize how far the various DNA fragments travel, the gel is soaked in ethidium bromide (EtBr) which binds to DNA and is fluorescent. When you expose the DNA to the correct wavelength of ultraviolet light, you can visualize your DNA fragments. See Fig. 14-11.

NUCLEAR PROBES

The base sequences on individual DNA fragments isolated by the Southern blot test can be determined using *nuclear* or DNA *probes*. Probes contain fragments of nucleotides. The single-stranded DNA in the electrophoresis gel will combine with strands of DNA containing base sequences that are complementary to the base sequences of the original strand. If you are looking for a base sequence of A-T-C-G-G, you would use a probe containing the sequence T-A-G-C-C. The resulting double strand is called a hybrid because it is a combination of the natural DNA and your "unnatural" probe. The hybridization step consists in simply applying the single-strand probes to the target DNA.

Hopefully you thought ahead and provided your probe with some kind of label. Your choices are a radioactive or a fluorescent label. One way to detect target molecules is with a system of coupled antibodies and fluorochromes, a method known as "fluorescent in-site hybridization" (FISH). The probes can be synthesized

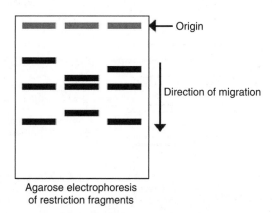

Agarose electrophoresis
of restriction fragments

Figure 14-11 Visualization of migration of DNA fragments using ethidium bromide.

with incorporated fluorescent molecules or molecules which can be recognized with fluorescent antibodies so that the direct visualization of the probes is possible.

In either case, you will dry the gel out on a filter system to provide a solid matrix for the analysis. If you have labeled your DNA fragments with a radioactive atom such as tritium, you will lay a photosensitive material over the matrix (film) and the radioactive material will expose this material just like light would. If you have labeled your DNA fragment with a fluorescent label, then you will expose the matrix to UV light of the appropriate wavelength. Obviously, you will need a label that requires a different wavelength of light than does EtBr. In any case, you should see something like this (Fig. 14-12).

From this analysis you learn the location of the DNA sequences specific for your restriction enzymes and the location of the base sequences complementary to your probe within the DNA fragment.

SANGER TECHNIQUE

The Sanger technique has been used since its development in 1977 for DNA sequencing. It is called the "chain terminator method" because it is a very clever method of determining the location of a specific base on a DNA fragment, based on where synthesis of a new DNA chain stops. The methodology depends on the fact that (1) synthesis of a double-stranded DNA strand from a single strand of DNA will be initiated in the presence of DNA polymerase, and (2) DNA synthesis will stop if the incorporated base is in the form of dideoxynucleotide instead of deoxynucleotide. The dideoxy form of the nucleotide is missing a hydroxyl group at a critical point.

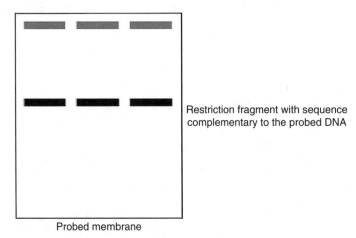

Restriction fragment with sequence complementary to the probed DNA

Probed membrane

Figure 14-12   Location of DNA probe.

So, if you provide a batch of synthesizing DNA molecule with, for example, dideoxynucleoadenosine (ddATP) in a mixture that also contains deoxynucleo-adenosine (dATP), as well as the other three deoxynucleotides, the synthesis of the double chain will stop when the ddATP molecule is incorporated instead of the dATP molecule. By the laws of probability, some of the synthesizing DNA will be stopped at every point that adenosine is required. See Fig. 14-13.

The Sanger technique uses the dideoxynucleotide for all four of the required nucleotides. There are four batches of reagents, one devoted to each nucleotide. The same single-stranded DNA molecule is incubated in each batch with one of the nucleotide provided in the dideoxy form as well as its normal deoxy form. Among the four batches, synthesis has been arrested at every site in the DNA fragment. You keep the batches separate and run gel electrophoresis on all four batches.

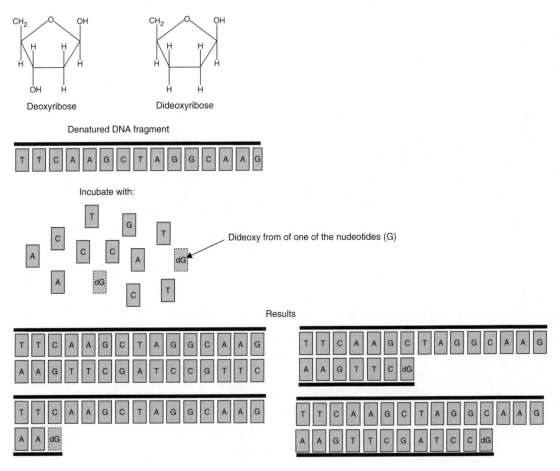

Figure 14-13   Use of dideoxynucleotide to terminate DNA synthesis.

Because the length of the migration in the electrophoresis field depends on the size of the molecule, the DNA fragments should distribute themselves in linear fashion according to size. See Fig. 14-14 to understand how, when you compare the four runs of gel electrophoresis, you can tell which base has been incorporate at every site in the DNA fragment.

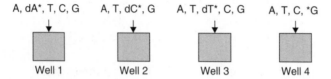

Template—Mystery fragment

$X_1, X_2, X_3, X_4, X_5, X_6, X_7, X_8, X_9$

Primers—Nucleotides, *Radioactive dideoxyribose from

A, dA*, T, C, G   A, T, dC*, G   A, T, dT*, C, G   A, T, C, *G

Well 1   Well 2   Well 3   Well 4

Gel electrophoresis with dideoxy nuclectides visualized with audioradioagraphy

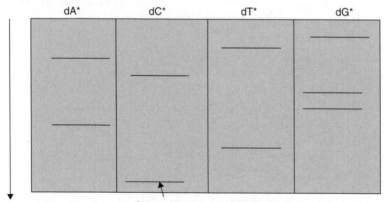

Shortest fragment—indicating first organic base in sequence

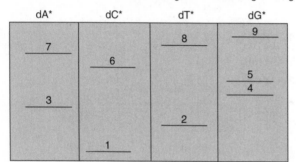

Primer sequence—CTAGGCATG

Mystery DNA sequence—CTACCCGTAC

Figure 14-14   Determining DNA sequence using the Sanger method.

DNA SYNTHESIZERS

The technique that you have just learned would enable you to analyze all bioactive proteins, if the living organism were not a complex system of checks and balances. However, most bioactive molecules are modified after their initial production from the mRNA. The modification requires other proteins that are produced according to other environmental signals and act as a way to fine-tune the action of the hormone. For example, insulin is produced as preproinsulin, is stored as proinsulin, and is then secreted as insulin. So if you produce a protein from the "insulin gene," you will produce preproinsulin. Modification of the pre-insulin molecules would require the enzymes that are responsible for producing the final form. Rather than attempt to use natural methods to produce the final protein, methods have been developed to work backwards from the amino acid structure of the protein of interest and develop a gene to code for that protein, even though such a gene never existed in nature. In other words, researchers can build artificial DNA that codes directly for a given protein.

You can synthesize DNA outside of a living cell because the chemical forces that bind the DNA together are operative inside or outside of the cell. To begin the synthesis of a new strand of DNA, you need a matrix to hold the molecules in place, thereby enabling their interactions. DNA synthesizers use a silica matrix that binds the first nucleotide in the sequence present in a small column. The second nucleotide is then flushed through the column and binds to the first nucleotide by means of the sugar-phosphate bond that forms the DNA backbone. The molecules of the second nucleotide have to be chemically blocked so that they will only bind to the first nucleotide and not to one another. The excess of the second nucleotide is flushed out of the column and the chemical blockage removed from the nucleotide that has bound to the matrix. Then the third nucleotide is added and so on.

This process is currently performed by a computerized system that, once provided with the desired DNA sequence, will produce the DNA strand. These systems are called *DNA synthesizers*.

INFORMATION RESOURCES

We are truly—no kidding—living in an information age. In generations past, new scientific developments were shared intermittently at scientific conferences between colleagues or by the occasional publication of the revolutionary scientific paper. Beginning in the 1980s, the scientific community devised ways to communicate small advances and tiny but crucial steps almost instantaneously using databases accessed through network connections.

A number of scientific institutions are maintaining databases to track and index genetic information as it is derived. These include the Online Mendelian Inheritance in Man (OMIM), a database maintained by researches at John Hopkins School of

Medicine. This database is devoted to human genes, genetic traits, and disorders. Locus link is a National Centre for Biotechnology Information (NCBI) database that serves as an interface to genetic information from a variety of bioinformatics sources. GeneCards was developed at the Weizmann Institute of Science in Israel and is a database of human genes and hereditary disorders. There are species specific databases. There are databases devoted to variability among given alleles. European molecular biology laboratory (EMBL), GenBank, and the DNA Database of Japan (DDBJ) record all available information of DNA sequences. The Protein sequence database (PSD) of the protein identification resource (PIR) contains almost 30,000 entries. Other databases devoted to protein sequence are the Munich Information Center for Protein Sequences (MIPS) and Swiss-Pro. There are also protein family databases that document results of analyses of sequences. Such databases match domains and contain information about homologous functions. Protein structure databases, such as PDB, contain the results of three-dimensional structures created by crystallographic and spectroscopic studies. Databases are available to search all the other databases. These and many other tools attempt to consolidate all of the massive information being generated at a daunting rate with the aim of facilitating further progress in the realm of human genetics.

Analytical Techniques in Deriving Relationships Between Genomes

Now that we have the blueprint of humans, we are turning our attention to other species. We hope to better understand our relationship, genetically speaking, to our brethren creatures. In some cases, we can clearly see how our genome has changed relative to other creatures. For example, examine Fig. 14-15. This figure shows chromosome 2 from humans compared to chromosome pairs from chimpanzees. The striping within the chromosomes is due to dense regions that stain darkly, called heterochromatin. The similarities in patterns suggests that the primate chromosomes fused somewhere along the evolutionary chain that resulted in humans.

Usually, the pattern is not as clear. The problem is partly due to the large fraction of noncoding introns that interrupt the codon sequence. Different species may have exactly the same gene but with different introns inserted. Another complexity is the huge percentage of the genome that does not code for proteins. In humans, this amounts to 95% of the DNA. Even though a segment of the DNA does not seem to produce a protein, an analogous segment on DNA of another species may indicate a relationship. Various DNAs are compared by techniques of *sequence alignment*.

For a very simple example, consider the following sequence:

Start   CAT   AGC   TAA   CTA   Stop

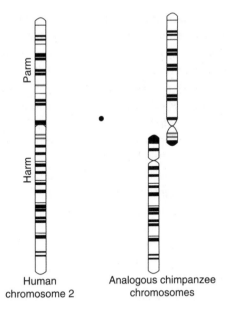

Figure 14-15   Schematic Showing Similarities in Banding Among Chromosomes from Different Species.

You might compare this with the following sequence:

Start   CAA   AGC   TAT   CTA   Stop

These comparisons reveal two *single nucleotide polymorphisms (SNPs)*, or differences of single bases on a specific spot on the genome, where CAT has become CAA and TAA has become TAT. There have been about 4 million SNPs identified among the few individual human genomes that we have. Incidentally, one of these was released in the year 2007 and is from the indefatigable Dr. Craig Venter. Dr. Venter is one of the pioneers of genomic research and was instrumental in the original sequence of the human genome. (See Genome Wars by James Shreeve).

Now let's compare our original sequence with the following:

CAT   TT   AGC   TAA   G   CTA

Remember that I put in the spaces for clarity. There are no spaces within the DNA. The above could lead one to conclude that extraneous material had been inserted. However, this sequence might cause a *frame shift*, where the code is read as

CAT   TTA   GCT   AAG   CTA

Now it looks completely different. Imagine the complexity when the comparative sequences consist of hundreds of bases.

Computer algorithms have been developed to help decipher relationships between sequences. A computer algorithm is simply a mathematical analysis directed toward a specific problem. All modeling and simulation problems are attacked using computer algorithms. Typically computer algorithms are much more complex that would be realistic for a human to use without computer aid. The computer algorithms that compare genetic sequences find the best possible alignment for different sequences and assign priorities to different indices of similarity. Such a program would align the sequence of interest with other sequences, seek areas that match and apply penalties for gaps. The researchers use the results from the computer analysis to conclude whether the similarities/differences are biologically meaningful.

The selection of the sequence to be studied depends on the goals of the study. If the interest is in establishing relationships across a broad range of species, then a gene segment related to a ubiquitous function would be selected, for example, genes coding for ribosome structures. If the interest is in deciphering relationships between closely related species, such as gorillas, chimpanzees, and humans, then a more esoteric function might be selected. If one is looking at relationships between groups of humans, the DNA on mitochondria and on the Y chromosome would be examined because these two entities pass unaltered between mothers and their children, and between fathers and their sons, respectively.

The outcome of a study of the similarities and differences between gene segments is a phylogenetic tree. The tree shows branch points where groups of species have diverged. These branch points are called *nodes*. The group of species that share a common node is called a *clade*. Fig. 14-16 shows a hypothetical phylogenetic tree. According to these results, species A, B, and C diverged from a common ancestor with D. Then A and B diverged from C. Such proposal are compared with similar trees developed on the basis of other computer algorithms (there are several) and on comparisons of other gene segments shared among the species in questions.

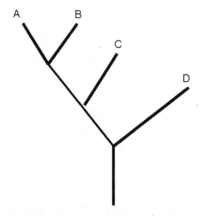

Figure 14-16   A phylogenetic tree.

Many of the conclusions from molecular evolution studies confirm traditional wisdom based on physiological similarities. An example is the close relationship between humans and chimpanzees. There have been surprises, however, such as the similarities between primates, rodents, and shrews. Like all scientific developments based on a new and revolutionary technology, we can expect some newsworthy findings over the next few decades.

THE HUMAN GENOME

It would not be right to leave a discussion on bioinformatics without addressing the current state of knowledge regarding the human genome. Admittedly, such knowledge is in its infancy. We have the human genome sequence almost completed. We are still missing some short segments in the noncoding regions. This sequence is based on a relatively few individuals, so we may learn more about human diversity in the years to come. What we have found is startling.

Of the approximately 3 billion base pairs on human DNA, only 1.5% codes for proteins. The rest was initially referred to as "junk." However, a closer look indicates that the "junk" may actually have an impact. For one thing, most of the genome appears to be transcribed, or readout by RNA. The impact of such transcription is a subject of ongoing research. It appears that the term "selfish" is appropriate for some of the DNA. That is to say that this material has inserted itself into our genome, perhaps from viral origin, and exists with the sole result that it is duplicated within the human cell.

The noncoding portion of the genome includes the following elements:

DNA transposons—2.9%   These are sequences of DNA that can move around in different parts of the genome. Transposable elements are a source of mutations and appear to account for at least some of the differences between humans and primates. In some plants, over half of the genomes consist of transposable elements.

Retrotransposons—41.8%   These elements are similar to retroviruses. They readout from the DNA as RNA. The enzyme reverse transcriptase creates the complementary DNA from the RNA and the cDNA is reinserted into the genome. Retrotransposons can amplify themselves in the genome. However, most appear to be ancient and no longer able to function. There are two types of retrotransposons—long terminal repeats (LTR) and non-LTR.

1. **Long terminal repeats (LTRs)—8.3%**   LTRs were first described in retroviruses where they were observed to initiate readout of the code. LTRs consist of long repeated elements, some of which are inverted, that appear at each end of the coding regions.

2. **Non-LTR**   The non-LTR can be divided into *long interspersed nuclear elements (LINEs)* and *short interspersed nuclear elements (SINEs)*.

- **LINEs—20.4%**   LINEs consists of two segments—one that binds to mRNA and the other codes for the accompanying reverse transcriptase and for endonucleases that insert complementary DNA into the genome. LINE elements can both copy themselves and also can copy SINEs.

- **SINEs—13.1%**   These are short DNA sequences that once were reverse-transcribed RNA molecules. SINEs are transposed in conjunction with LINEs because they don't have their own reverse transcriptase. Primates contain a very interesting and unique type of SINE *ALU sequences*. ALU elements were named so because they are targeted by the restriction enzyme ALU. The movement of ALU elements appears to have played a significant role in evolution as well as in human genetic disease and cancers.

Simple sequence repeats—3%   Parts of the genome appear to stutter, repeating the same short base sequences over and over. One piece of evidence that indicates the importance of the simple sequence repeats is the fact that Huntington's chorea is caused by a mutation within one of these genetic stutters.

Segmental duplication—5%   As the name implies, these are areas where segments of the genome are repeated.

Miscellaneous unique sequences—11.6%   These are sequences that do not repeat. Function is unknown.

Miscellaneous heterochromatin—8%   The heterochromatin stains darkly and creates the banding patterns seen in condensed chromosomes. The dense regions of telomeres, capping the ends of eukaryotic chromosomes, and centromeres, at the center, are heterochromatin. Sequencing of heterochromatin is incomplete.

The preliminary conclusions about the human genome smack more of philosophy than science. We share 99.9% of our genome with other human beings. There appears to be no genetic bases for race. We share 95% of our genome with rodents. One can hope that the knowledge of our close relationships will motivate us to treat one another and our fellow creatures more decently.

Summary

The science of bioinformatics utilizes computer technology to provide analysis of the massive amount of data generated by the sequencing of DNA from various organisms and by the analysis of protein structure and function. One of the most important developments associated with bioinformatics is the creation of massive databases, freely available to the scientific community and searchable by available search tools.

DNA contains the formula for amino acids in the form of three base codons. A segment of DNA with a start and stop signal constitutes a gene—with the correct sequence of amino acids for a given protein. The production of proteins begins with the attachment of RNA polymerase to a segment of the DNA. The RNA polymerase recognizes a specific configuration of the DNA produced by binding to the DNA of an activation factor, such as a hormone. The RNA polymerase creates a single-stranded RNA molecule, called mRNA, which is composed of organic base sequences that are complementary to the sequence on the DNA. The mRNA contains some extraneous regions, called introns, which must be edited out. The mRNA leaves the nucleus and binds to ribosomes. The ribosomes rachet along the RNA strand and assemble the protein by recruiting transfer RNA that is bound to the correct amino acid.

The repertoire of proteins within a cell is called the *proteome*. Standard techniques are used to isolate and characterize a protein from among a given cell's proteome. These include methods to separate molecules on the basis of size, including gel electrophoresis and chromatography. Both gel electrophoresis can be modified to also separate molecules on the basis of chemical characteristics, such as charge. The three-dimensional structure of proteins can be visualized using x-ray spectrophotometry and protein nuclear magnetic resonance.

DNA sequencing is possible only because of the polymerase chain reaction (PCR) technology that allows production of large quantities of a given DNA sequence from a small fragment. DNA sequencing is done by automation of a long-standing technique called the Southern blot test. This test starts with the fragmentation of DNA by restriction enzymes. These are enzymes which cut the DNA at known points, depending on the specific restriction enzyme used. The fragments are then submitted to gel electrophoresis and separated on the basis of size. This technique is coupled with Sanger technique whereby the synthesis of a double-stranded DNA strand from a single strand of DNA is initiated in the presence of DNA polymerase, and synthesis is terminated by incubating the sample with one of the incorporated bases in the form of dideoxynucleotide instead of deoxynucleotide. This technique generates DNA fragments of different lengths dependent on where the altered base is incorporated. By using different samples incubated with each of the four organic bases in the dideoxynucleotide form, the synthesis of the fragment is stopped at every point. When the samples are submitted to gel electrophoresis, they sort according to size, revealing where bases are located. This process is automated, and massive amounts of data are being generated regarding DNA sequences.

DNA sequence analysis is nontrivial and relies on computer algorithms to determine the similarities and differences between sequences that have mutated and moved within different genomes. This research has revealed fascinating relationships between organisms from a vantage point called molecular evolution. In many cases, the results from molecular evolution duplicate traditional wisdom based on physical similarities. In other cases, relations that we thought were distant, such as between humans and rodents, have been found to be quite close.

The human genome has yielded startling information. Less than 2% of it codes for proteins. The rest is duplicating, stuttering, and potentially viral in origin. The most important fact to date is that we humans share 99.9% of our genome in common.

Quiz

1. The shape of a protein can be discerned by

   (a) x-ray crystallography.

   (b) enzyme-linked immunosorbent assay (ELISA).

   (c) gel electrophoresis.

   (d) column chromatography.

2. The results of mass spectrophotometry can be used to determine

   (a) size.

   (b) shape.

   (c) ionic charge.

   (d) identity of the molecule based on size/charge ratio.

   (e) All of the above.

3. The mRNA for a given protein can be eliminated from a cell by

   (a) antisense technology.

   (b) monoclonal antibodies.

   (c) microarray.

   (d) All of the above.

4. The proteins being actively produced by a cell can be determined by

   (a) mass spectrophotometry.

   (b) protein NMR.

   (c) DNA microarray.

   (d) x-ray crystallography.

5. Polymerase chain reaction is important because

   (a) it provides many copies of a given fragment of DNA.

   (b) it reveals the sequence of amino acids on the DNA.

   (c) it reveals which sections of the DNA are being expressed.

   (d) it provides the amino acid sequence for a given protein.

6. The proteome differs from the phenome because

   (a) the proteome is an expression of the genome.

   (b) the phenome is a cell characteristics, while the proteome is the code for a cell characteristics.

   (c) the proteome is only the protein part of the phenome.

   (d) All of the above are correct.

7. In transcription, the ribosome serves to

   (a) interpret the DNA code.

   (b) have no role in transcription.

   (c) recruit the appropriate amino acids for assembly of the protein.

   (d) interpret the mRNA.

8. The Sanger technique relies on the fact that

   (a) DNA synthesis stops if the organic base is in the form of a dideoxyribose nucleic acid.

   (b) restriction enzymes cut DNA at predictable points.

   (c) different DNA fragments stain differently.

   (d) reverse transcriptase will cause the formation of DNA from mRNA.

   (e) All of the above.

9. Elements that can move around the genome are called

   (a) LINEs.

   (b) SINEs.

   (c) DNA transposons.

   (d) retrotransposons.

   (e) All of the above.

10. The majority of the human genome

   (a) are short segment repeats.

   (b) are unique to each individual.

   (c) do not code for proteins.

   (d) remain unknown.

   (e) consist of long interspersed nuclear elements.

Final Exam

1. Hydrochloric acid dissociates in water to form a positive hydrogen ion and a negative chlorine ion. The type of bond represented by hydrochloric acid is

   (a) ionic.

   (b) covalent.

   (c) hydrogen.

   (d) thermodynamically unstable.

   (e) Both (a) and (d) are correct.

2. Entropy

   (a) is a phenomenon that governs all systems in the universe.

   (b) drives a system and its surroundings to an overall state of greater randomness.

   (c) takes energy to overcome for systems which build increased order.

   (d) is related to the temperature of a system.

   (e) All of the above are correct.

3. A solution that has a pH of 10.0

   (a) is acidic.

   (b) is basic.

   (c) has a large concentration of hydrogen atoms compared to water.

   (d) has a hydrogen ion concentration of 0.01 mol.

   (e) (a) and (c) are correct.

   (f) Both (b) and (d) are correct.

4. An increase in entropy for a reaction

   (a) increases the change in free energy for the reaction.

   (b) decreases the change in free energy for the reaction.

   (c) means that more energy is invested in maintaining order.

   (d) requires less energy at high temperatures.

   (e) Both (a) and (c) are correct.

5. An equilibrium constant is a measure of

   (a) the rate of formation of product in a reaction.

   (b) the relative amount of the product versus reactant that exists in a system in equilibrium.

   (c) the amount of free hydrogen ion.

   (d) the entropy content of the system.

6. The presence of a buffer will

   (a) increase the change in free energy of a reaction.

   (b) buffer the change in free energy of a reaction.

   (c) reduce the change in pH for a reaction that releases or absorbs hydrogen ions.

   (d) shift the equilibrium constant for a reaction that releases or absorbs hydrogen ions.

7. The oxidation of carbon

   (a) causes the valence electrons to move to a lower energy state.

   (b) releases energy.

   (c) produces reduced forms of hydrogen.

   (d) drives the process of life.

   (e) All are correct.

8. We know that bacteria and archaea are closely related because

    (a) they share RNA sequences that eukaryotic cells do not have.

    (b) they both are prokaryotic.

    (c) they both have a cell wall.

    (d) All of the above are correct.

    (e) None of the above are correct.

9. The cell membrane forms a bilayer because

    (a) the lipid portions of the molecules cluster together.

    (b) the polar portions of the molecules are attracted to one another.

    (c) the exterior of the cell is hypertonic to the interior, forcing the membrane to close.

    (d) cell processes actively fold the membrane into a circle.

    (e) Both (a) and (b) are correct.

10. The large surface area of many cell membranes reflects the fact that

    (a) endocytosis has resulted in the intake of a large portion of the cell membrane.

    (b) important metabolic processes occur across membrane surfaces.

    (c) membranes help to segregate chemical processes.

    (d) membranes help to regulate the influx and efflux of water.

11. Mitochondria

    (a) are found only in eukaryotic organisms.

    (b) provide an isolated sac where an electrochemical gradient can exist across a membrane.

    (c) provide surface area for reactive biomolecules.

    (d) Both (b) and (c) are correct.

    (e) All of the above are correct.

12. Gram-negative bacteria

    (a) do not have a cell wall.

    (b) are susceptible to penicillin because they have a cell wall.

    (c) are archaea.

    (d) are found in extreme environments.

    (e) None of the above are correct.

13. The rough endoplasmic reticulum

    (a) has its own DNA.

    (b) contains digestive enzymes.

    (c) provides surface area for metabolic processes.

    (d) All of the above are correct.

    (e) None of the above are correct.

14. Mitochondrial DNA is useful in determining relationships between groups because

    (a) it is single stranded.

    (b) it is inherited strictly from the mother.

    (c) it does not contain mutations.

    (d) it contains only four bases.

    (e) All of the above are correct.

15. Lysosomes

    (a) are specialized peroxisomes.

    (b) are capable of digesting lipids.

    (c) concentrate enzymes for oxidative metabolism.

    (d) grow and divide on their own.

    (e) All of the above are correct.

16. The location of ribosomes on the endoplasmic reticulum surface

    (a) brings the two ribosomal subunits together.

    (b) provides a membrane for the establishment of a proton gradient.

    (c) provides a membrane to embed the transfer RNA.

    (d) locates newly produced proteins close to areas where they undergo further processing.

    (e) All of the above are correct.

17. Enzymes can be best described as

    (a) proteins that catalyze biochemical reactions by altering substrate concentrations.

    (b) proteins that catalyze biochemical reactions by increasing the rate of a reaction.

(c) proteins that enable biochemical reactions to proceed at a rate suitable for the cellular environment by decreasing the rate of a reaction.

(d) as long carbon chain molecules that catalyze biochemical reactions by increasing the rate of a reaction.

18. If the degree of inhibition of an enzyme is unaffected by the concentration of substrate we say that

(a) competitive inhibition exists.

(b) anti-competitive inhibition exists.

(c) pure noncompetitive inhibition exists.

(d) uncompetitive inhibition exists.

19. The four types of enzyme inhibition can be best classified as

(a) pure noncompetitive, uncompetitive, pure competitive, and mixed.

(b) There are only three types of inhibition: noncompetitive, anticompetitive, pure competitive.

(c) pure noncompetitive, pure anticompetitive, mixed noncompetitive, and mixed anticompetitive.

(d) There are only two types of inhibition: noncompetitive and competitive.

20. The induced fit model is best described by which of the following statements?

(a) As a substrate begins to bind to an enzyme, the enzyme induces a conformational change in the structure of the substrate allowing it to tightly bind to the enzyme.

(b) The presence of a substrate in solution causes a conformational change in the shape of an enzyme.

(c) Enzymes are rigid structures, so the induced fit model has been discredited.

(d) The binding of a substrate to an enzyme causes the enzyme to change shape, tightly binding the substrate.

21. One way that enzymes can act to catalyze a reaction is by

(a) lowering the activation energy for the reaction.

(b) lowering the standard reduction potential for a reaction.

(c) raising the standard reduction potential for a reaction.

(d) raising the activation energy for a reaction.

(e) lowering the binding energy for a reaction.

22. A *cofactor* can be best described as
    (a) an inorganic compound that inhibits substrate binding.
    (b) an inorganic compound that promotes substrate binding.
    (c) an inorganic or organic compound that activates an enzyme.
    (d) an organic substrate that promotes the binding of an additional substrate.

23. An *apoenzyme* is
    (a) an enzyme that does not require a cofactor.
    (b) an enzyme that requires a *bound* cofactor.
    (c) an enzyme that requires a cofactor, which is *not* bound.
    (d) an enzyme that requires an *organic* cofactor.

24. Which statement about coenzymes is not true?
    (a) Coenzymes are called vitamins when they must be acquired from the diet.
    (b) Coenzymes must be acquired from the diet.
    (c) Coenzymes carry important functional groups like acetyl groups.
    (d) Coenzymes are not required by all enzymes.

25. In competitive inhibition
    (a) a chemical called an *inhibitor* competes with the substrate for a binding site on the enzyme.
    (b) a chemical called an *inhibitor* can bind to a substrate preventing the enzyme from doing so.
    (c) a chemical called an *inhibitor* binds to the enzyme-substrate complex, preventing the reaction from going forward.
    (d) a chemical called an *inhibitor* binds to a *product* molecule after an enzyme mediates reaction, and reverses it.

26. A *hydrolase* is an enzyme which
    (a) cannot function in the presence of H_2O.
    (b) requires H_2O to induce a change in the spatial arrangement of a molecule.
    (c) releases H_2O after cleaving C—C, C—N, C—O, and other bonds.
    (d) requires H_2O to cleave C—C, C—N, C—O, and other bonds.

27. Which of the following statements about glycolysis is *not* true?

   (a) Compounds other than glucose can enter the glycolysis pathway at intermediate stages.

   (b) Glucose is converted to succinate, which is oxidized in the citric acid cycle.

   (c) For each molecule of glucose that is consumed, 2 ATP molecules are produced.

   (d) Glucose is converted to pyruvate, which can be oxidized in the citric acid cycle.

28. In the initial step of glycolysis

   (a) glucose is converted into iso-glucose.

   (b) glucose is transformed into a form which is more easily oxidized.

   (c) glucose is converted into glucose 6-phosphate.

   (d) glucose is converted into glucose 3-phosphate.

29. Which of the following is the best characterization of the change of free energy in glycolysis?

   (a) Some steps are exergonic and some are endergonic, but the overall process is exergonic with more than enough energy to produce several molecules of ATP.

   (b) Some steps are exergonic and some are endergonic, but the overall process is exergonic with barely enough energy left over to produce 2 molecules of ATP.

   (c) All steps in the reaction are exergonic.

   (d) All steps in the reaction, except the conversion of phosphoenolpyruvate to pyruvate are endergonic.

30. The enzyme *glucokinase*

   (a) is found only in the liver, and is inhibited by glucose 6-phosphate.

   (b) is found only in the skeletal muscle, and is inhibited by glucose 6-phosphate.

   (c) is found only in the skeletal muscle, and is not inhibited by glucose 6-phosphate.

   (d) is found only in the liver, and is not inhibited by glucose 6-phosphate.

31. The enzyme pyruvate kinase

   (a) is activated by fructose, and inhibited by ATP.

   (b) is activated by ATP, and inhibited by fructose 1,6-biphosphate.

(c) is activated by fructose 1,6-biphosphate and is inhibited by ATP.

(d) is inhibited by fructose 1,6-biphosphate.

32. Phosphofructokinase is

(a) activated by low levels of ATP.

(b) not involved in glycolysis.

(c) activated by high levels of ATP.

(d) inhibited by ATP.

33. While heart muscles contain a predominance of the H4 isozyme and skeletal muscles contain a predominance of the M4 isozyme, liver cells

(a) contain a predominance of H4, explaining the low affinity for pyruvate in liver cells.

(b) contain a predominance of M4, explaining the low affinity for pyruvate in liver cells.

(c) contain a predominance of M4, explaining the high affinity for pyruvate in liver cells.

(d) contain neither M4 nor H4 isozymes.

34. Which of the following statements is not true?

(a) Pyruvate is converted to ethanol by yeast cells.

(b) Pyruvate is converted to ethanol using pyruvate decarboxylase and alcohol dehydrogenase.

(c) Pyruvate can be converted to ethanol in human muscle cells, but only under very restrictive conditions.

(d) Pyruvate cannot be converted to ethanol in mammalian cells.

35. The activity of phosphofructokinase

(a) is enhanced by ADP or AMP and inhibited by ATP and NADH.

(b) is enhanced by ADP or AMP and inhibited only by ATP.

(c) is inhibited by ADP and enhanced by ATP.

(d) is inhibited by ADP and enhanced by NADH.

36. Mannose can enter the glycolysis pathway when

(a) it is converted to glucose 6-phosphate.

(b) it cannot enter the glycolysis pathway.

(c) it is reduced to simple glucose.

(d) it is converted to fructose 6-phosphate.

37. The backbone of a protein is able to form hydrogen bonds between

    (a) hydrophobic side chains and hydrophilic side chains.

    (b) hydrogen bond acceptor and receivers on the alpha carbon.

    (c) cysteine residues.

    (d) electron-rich carboxyl groups.

    (e) All of the above are correct.

    (f) Both (b) and (d) are correct.

38. In a folded protein, one would expect

    (a) hydrophilic groups on the outside where they interact with their environment.

    (b) hydrophobic groups on the inside where they stabilize the protein by ionic bonds.

    (c) both hydrophilic and hydrophobic groups on the inside forming hydrogen bonds.

    (d) hydrophilic groups on the inside where they stabilize the protein by ionic bonds.

39. A protein that is intended to be secreted through the cell membrane

    (a) would probably consist entirely of hydrophilic side chains.

    (b) would probably require refolding to place hydrophobic groups on the outside.

    (c) would probably be refolded to place hydrophilic groups on the outside.

    (d) would probably consist entirely of hydrophobic side chains.

40. Amino acids with uncharged polar side chains

    (a) participate in covalent bonds in the secondary and tertiary structure of proteins.

    (b) participate in hydrogen bonds in the secondary and tertiary structure of proteins.

    (c) have entities that are charged at physiological pH, such as carboxyl groups.

    (d) are essential amino acids.

41. Serine is often found at the active site of enzymes because

    (a) it has the ability to accept or donate a protein, depending upon the pH of the cell microenvironment.

    (b) it is charged at physiological pH and therefore is found on the outside of proteins.

(c) it is very small and able to fit into tight configurations.

(d) it forms a kink in the secondary structure.

(e) Both (c) and (d) are correct.

42. Asparagine is important in the ability to

(a) bind substrate to enzymes.

(b) perform neurotransmission.

(c) make glycoproteins.

(d) form an alpha helix.

43. The imidazole ring of histidine

(a) is uncharged at physiological pH.

(b) behaves like a lipid.

(c) is active in proton exchange.

(d) is found at the active site of enzymes.

(e) can form hydrogen bonds.

(f) (c), (d), and (e) are all correct.

(g) Both (a) and (b) are correct.

44. Collagen is formed by

(a) beta sheets.

(b) alpha helices.

(c) unique three-stranded polypeptide structures.

(d) a combination of alpha helices and beta sheets.

45. In proteins, domains are

(a) where alpha helices connect to beta sheets.

(b) structural modules found in a number of different proteins.

(c) areas of electromagnetic activity.

(d) the portion of a molecule actively involved in interfaces with the environment.

46. The fixation of nitrogen requires

(a) electrons.

(b) protons.

(c) energy.

(d) Mo-Fe protein.

(e) All of the above are correct.

47. Nitrogen fixation

    (a) requires oxygen.

    (b) releases energy.

    (c) can be done by most plants.

    (d) occurs when lightening strikes.

    (e) None of the above are correct.

48. The production of nonessential amino acids in humans

    (a) requires transaminases.

    (b) requires glutamine usually.

    (c) depends on the availability of the corresponding α-keto acid.

    (d) is tied to key metabolites of gluconeogenesis.

    (e) All of the above are correct.

    (f) All except (d) are correct.

49. Transport of ammonia to the kidney

    (a) requires Mo-Fe protein.

    (b) requires transformation of the ammonia to a nontoxic form.

    (c) can be done by conversion of glutamate to glutamine by glutamine synthetase.

    (d) can be done by conversion of glutamate to α-ketoglutarate by glutamate dehydrogenase.

    (e) Both (b) and (c) are correct.

50. Biosynthesis of glutamate

    (a) can be done by the action of glutamine dehydrogenase on α-ketoglutarate.

    (b) can be done by the action of glutamine synthase on α-ketoglutarate.

    (c) can be done by the action of glutamine synthetase on glutamine.

    (d) All of the above are correct.

    (e) Both (a) and (b) are correct.

51. The formation of a single molecule of urea requires the cleavage of

    (a) one molecule of ATP.

    (b) three molecules of ATP.

    (c) four molecules of ATP.

    (d) six molecules of ATP.

52. The role of aspartate in the urea cycle is best described as

    (a) Serves as an intermediate.

    (b) Donates an amine to α-keto glutarate.

    (c) Donates an amine to the final product.

    (d) Provides energy.

53. Phenylketonuria

    (a) is due to insufficient phenylalanine in the diet.

    (b) is due to inability to correctly incorporate nitrogen into the α-keto acid precursor to phenylalanine.

    (c) is due to an inability to produce tyrosine.

    (d) is an example of starvation in the face of adequate caloric intake.

54. Sulfonamide drugs are not toxic to humans as they are to bacteria because

    (a) sulfonamides attack the cell wall.

    (b) humans break down the sulfonamide in the gastrointestinal (GI) tract.

    (c) sulfonamides interfere with the production of essential amino acids in bacteria, and humans consume essential amino acids in their diet.

    (d) humans ingest folic acid in their diet.

55. Endocrine glands

    (a) release hormones that influence cells in the nearby vicinity.

    (b) release hormones that influence distant cells.

    (c) release hormones that influence cells by diffusion.

    (d) produce only steroid hormones.

56. Steroids are compounds that can be described as

    (a) hormones metabolized into cholesterol.

    (b) synthetic compounds that resemble naturally occurring substances.

    (c) hormones derived from estrogen.

    (d) hormones derived from cholesterol.

57. An androstane

    (a) is a synthetic steroid.

    (b) is a steroid containing 19 carbon atoms.

    (c) is a steroid only occurring in males.

    (d) is a steroid containing 20 carbon atoms.

58. Progesterone

    (a) is a steroid hormone secreted primarily by the corpus luteum.

    (b) is a steroid hormone secreted only by the corpus luteum.

    (c) is a steroid hormone not found in females during pregnancy.

    (d) is the only steroid hormone not derived from cholesterol.

59. Vitamin D

    (a) is sometimes confused with hormones due to a similar structure but is a vitamin.

    (b) can only be obtained from the diet.

    (c) comes in a single variety.

    (d) is a steroid hormone that must be obtained through diet or sunlight.

60. Vitamin D

    (a) can be immediately utilized by the body to increase calcium absorption.

    (b) must be converted to a biologically active form in a two-step process that takes place in the liver and kidneys.

    (c) must be converted to a biologically active form in a two-step process that takes place in the liver.

    (d) must be converted to a biologically active form in a two-step process that takes place in the kidneys.

61. Steroid hormones can act by

    (a) binding to cell membranes, changing the cell membrane potential.

    (b) transporting cofactors into the cell.

    (c) binding to a DNA site promoting the transcription of a gene.

    (d) binding to an RNA complex promoting the transcription of a gene.

62. Peptide Hormones

    (a) are short chains of amino acids that typically act as autocrine hormones.

    (b) are hormones which influence peptide synthesis.

    (c) are hormones that require a peptide cofactor.

    (d) are short chains of amino acids that typically act as endocrine hormones.

63. Epinephrine

  (a) is a neurotransmitter that increases heart rate and blood pressure.

  (b) is a neurotransmitter that reduces heart rate and blood pressure.

  (c) is a neurotransmitter that only increases contraction of skeletal muscle.

  (d) is not found in the brain.

64. Phenothiazine is effective in the treatment of schizophrenia because

  (a) it boosts dopamine levels in the brain.

  (b) it blocks the action of monoamine oxidase, enhancing dopamine's action in the brain.

  (c) it blocks D2 receptors in the brain, mitigating the effect of high levels of dopamine.

  (d) it enhances the action of monoamine oxidase, mitigating the effect of high levels of dopamine.

65. A reuptake inhibitor

  (a) is a drug which treats schizophrenia by blocking reuptake enzymes.

  (b) is a drug which treats Parkinson's disease by preventing reuptake of dopamine in the synapse.

  (c) is a drug which treats depression by inhibiting reuptake of norepinephrine and serotonin.

  (d) is a drug which treats depression by preventing reuptake of norepinephrine and serotonin.

66. The citric acid cycle

  (a) uses eight enzymes and the final product is malate.

  (b) uses eight enzymes and the final product is oxaloacetate.

  (c) results in the production of 1 ATP molecule.

  (d) results in the production of 2 ATP molecules.

67. The ingredient which feeds the citric acid cycle is

  (a) acetyl-CoA synthesized in the cytoplasm.

  (b) oxaloacetate synthesized in the cytoplasm.

  (c) oxaloacetate synthesized in the mitochondria.

  (d) acetyl-CoA synthesized inside mitochondria.

68. The carbon atoms of acetyl-CoA

 (a) are lost as CO_2 molecules.

 (b) are recycled with oxaloacetate.

 (c) play no role in the reaction.

 (d) are donated to succinate dehydrogenase.

69. In the citric acid cycle, the conversion of citrate to isocitrate

 (a) is done to make the molecule more suitable for reduction.

 (b) is slightly endergonic.

 (c) is done to make the molecule more suitable for oxidation.

 (d) is heavily exergonic.

70. The enzymatic activity of citrate synthase can be described as

 (a) an induced fit process.

 (b) as an uncompetitive inhibition process.

 (c) as a competitive inhibition process.

 (d) as a lock and key process.

71. Isocitrate dehydrogenase is considered a regulatory enzyme in the citric acid cycle because

 (a) its activity is inhibited by the presence of citrate.

 (b) its activity is inhibited by the presence of ADP.

 (c) its activity is inhibited by the presence of high-energy compounds.

 (d) its activity is regulated by cofactors produced in the cycle.

72. Succinate dehydrogenase is a regulatory enzyme in the citric acid cycle primarily because

 (a) its activity is inhibited by an accumulation of oxaloacetate.

 (b) its activity is inhibited by an accumulation of malate.

 (c) its activity is inhibited by high-energy compounds.

 (d) its activity is inhibited by an accumulation of citrate.

73. Cataplerotic reactions are beneficial because

 (a) they catalyze the production of an extra GTP molecule.

 (b) they catalyze the production of an extra NADH molecule.

 (c) they utilize intermediate compounds in the citric acid cycle, preventing their buildup in the mitochondria.

 (d) they extract extra energy from the citric acid cycle.

74. The electron transport chain

    (a) generates 5 ATP molecules.

    (b) prepares a proton gradient which makes ATP production possible.

    (c) utilizes large amounts of oxygen, so is damaging to the cells.

    (d) establishes an electron gradient, making ATP production possible.

75. The energy required for the synthesis of ATP in oxidative phosphorylation comes from

    (a) exergonic reactions in the electron transport chain.

    (b) oxygen in the mitochondria.

    (c) exergonic processes in the mitochondria.

76. Which of the following statements about complex I is not true?

    (a) Complex I contributes to the proton gradient.

    (b) Complex I catalyzes an exergonic reaction that transfers electrons from NADH to Coenzyme Q.

    (c) Complex I catalyzes the transfer of electrons from NADH to cytochrome c.

    (d) Complex I is bound to the inner mitochondrial membrane.

77. Which of the following statements about complex II is true?

    (a) Complex II catalyzes an exergonic reaction, but the energy is not sufficient to produce an ATP molecule.

    (b) Complex II catalyzes an exergonic reaction, but the energy is barely sufficient to produce an ATP molecule.

    (c) Complex II catalyzes an endergonic reaction.

    (d) Complex II contributes to the proton gradient.

78. In the electron transport chain-oxidative phosphorylation process, oxidation of NADH

    (a) does not contribute to ATP synthesis.

    (b) liberates enough energy for the production of 2 ATP molecules.

    (c) liberates enough energy for the production of 3 ATP molecules.

    (d) liberates enough energy for the production of 6 ATP molecules.

79. In oxidative phosphorylation, proton transport

    (a) helps move inorganic phosphate across the mitochondrial barrier.

    (b) contributes to the formation of ATP by inducing a conformational change in complex V.

(c) contributes to the formation of ATP by inducing a conformational change in complex IV.

      (d) contributes to the formation of ATP by inducing a conformational change in the γ chains.

80. Sphingolipids

      (a) are constructed from glycerol.

      (b) are only found in the nervous system.

      (c) are constructed from long-chain hydroxylated bases rather than glycerol.

      (d) contain at least two amine groups.

81. The effect of a single cis double bond in a long-chain hydrocarbon molecule is to

      (a) do nothing; it takes multiple cis bonds to affect the structure of the molecule.

      (b) cause a bend in the structure.

      (c) make the molecule more suitable to pack into a crystal.

      (d) make the molecule less suitable to pack into a crystal.

      (e) Both (b) and (d) are correct.

82. Why do fatty acids containing the same number of carbon atoms have different melting points?

      (a) Molecules without cis or trans double bonds pack together more closely giving crystals a higher melting point.

      (b) Molecules without cis double bonds pack together more loosely giving crystals a higher melting point.

      (c) Molecules without cis or trans double bonds remain straight chained, giving crystals a lower melting point.

      (d) Straight molecules require less energy to separate the molecules when heated.

83. The best description of how DNA and RNA are different is:

      (a) RNA is unable to form helical structures.

      (b) RNA uses ribose, while DNA uses deoxyribose.

      (c) RNA uses ribose, while DNA uses deoxyribose. In addition, RNA uses uracil while DNA uses thymine.

      (d) RNA does not have hydrophobic centers.

84. High-density lipoprotein (HDL)

    (a) only transports cholesterol in the bloodstream.

    (b) transports cholesterol to the kidneys.

    (c) releases cholesterol from the liver.

    (d) transports cholesterol from the tissues to the liver.

85. Hemoglobin binds oxygen by

    (a) oxidation of Fe^{2+} to Fe^{3+}

    (b) binding to 2,3-BPG (2,3-biphosphoglycerate)

    (c) partially oxidizing Fe and forming a cage with pyrrole subunits.

    (d) forming a cage with α helices.

    (e) None of the above are correct.

86. 2,3-BPG affects adaptation to high altitude by

    (a) increasing hemoglobin's ability to absorb oxygen from the atmosphere.

    (b) improving oxygen delivery to the tissues.

    (c) causing dilation of capillaries.

    (d) All of the above are correct.

87. A *left shift* in the oxyhemoglobin dissociation curve means

    (a) improved ability to absorb oxygen at low oxygen pressures.

    (b) improved distribution of oxygen to the tissues.

    (c) stabilization of the "t" configuration.

    (d) All of the above are correct.

88. Fetal hemoglobin

    (a) exhibits increased binding of BPG.

    (b) is similar to myoglobin.

    (c) is able to bind oxygen at low oxygen pressures.

    (d) All of the above are correct.

89. Ancillary pigments

    (a) are found in RCs.

    (b) form PSI and PSII.

    (c) remove the excited electron from activated chlorophyll.

(d) may absorb light in the green spectrum.

(e) None of the above are correct.

90. The PSI and PSII systems

    (a) are differentiated by the energy frequency preferred to activate an electron.

    (b) are differentiated by their end product.

    (c) are interconnected.

    (d) All of the above are correct.

91. Eukaryotes and prokaryotes are different in that

    (a) prokaryotes lack a thykaloid structure and therefore cannot establish a proton gradient.

    (b) prokaryotes don't have RCs.

    (c) prokaryotes are not subject to photorespiration.

    (d) All of the above are correct.

    (e) None of the above are correct.

92. The oxygen evolved during photosynthesis

    (a) originates with carbon dioxide.

    (b) is not evolved by sulfur metabolizing bacteria.

    (c) requires photorespiration.

    (d) All of the above are correct.

    (e) None of the above are correct.

93. The 48 photons of energy captured by photosynthesis

    (a) are captured with about 35% efficiency.

    (b) are used to fuel photorespiration.

    (c) are equal to the energy in one molecule of glucose.

    (d) are nearly equal to the energy in 18 molecules of ATP.

94. The ionic charge on a molecule can be used to separate the molecules in

    (a) chromatography.

    (b) gel electrophoresis.

    (c) mass spectrophotometry.

    (d) All of the above are correct.

    (e) Both (a) and (b) are correct.

95. Tandem mass spectrophotometry is used to determine

    (a) amino acid composition of a peptide.

    (b) sequence of amino acids on a peptide.

    (c) shape of a protein.

    (d) presence of a protein within a cell.

96. An antisense molecule

    (a) scrambles the amino acid sequence for a given protein.

    (b) binds to the DNA and prevents readout.

    (c) binds to the complementary mRNA strand.

    (d) binds to the protein and prevents function.

97. DNA microarrays

    (a) reveal which genes are present in a given genome.

    (b) reveal which proteins are being actively produced.

    (c) reveal where genes are located on the chromosome.

    (d) reveal which DNA sequences are functional.

98. The availability of the Taq polymerase is important because

    (a) it can produce DNA from RNA.

    (b) Taq can survive in the environment of hot springs.

    (c) Taq can survive the temperatures necessary for PCR technology.

    (d) Taq will attach at specific segments within the DNA.

99. Restriction enzymes

    (a) are found in bacteria.

    (b) restrict the production of proteins to only those needed by the cell.

    (c) restrict the production of proteins to ribosomes.

    (d) limit the uncoiling of DNA to the portions related to ongoing readout.

100. Protein function can be deduced using

    (a) DNA microarrays.

    (b) antisense technology.

    (c) analogy with proteins of similar structure.

    (d) conjecture based on amino acid sequence.

    (e) All of the above are correct.

Answers to Quiz and Exam Questions

Chapter 1

1. d
2. c
3. d
4. a
5. c
6. c
7. e
8. e
9. f
10. e

Chapter 2

1. d	6. a
2. d	7. c
3. c	8. e
4. c	9. e
5. b	10. g

Chapter 3

1. b	6. a
2. c	7. d
3. a	8. c
4. d	9. d
5. b	10. c

Chapter 4

1. c	6. d
2. c	7. b
3. e	8. c
4. b	9. a
5. c	10. a

Chapter 5

1. b
2. c
3. b

Chapter 6

1. d	6. a
2. a	7. c
3. b	8. c
4. b	9. b
5. a	10. d

Chapter 7

1. a	6. d
2. c	7. b
3. b	8. d
4. b	9. b
5. e	10. a

Chapter 8

1. c	6. a
2. c	7. b
3. b	8. a
4. a	9. c
5. c	10. b

Chapter 9

1. b	6. c
2. d	7. c
3. c	8. a
4. b	9. a
5. a	10. b

Chapter 10

1.	c	6.	b
2.	b	7.	c
3.	a	8.	b
4.	b	9.	d
5.	a	10.	a

Chapter 11

1.	c	6.	b
2.	b	7.	a
3.	c	8.	b
4.	a	9.	c
5.	d	10.	a

Chapter 12

1.	b	6.	c
2.	c	7.	b
3.	a	8.	b
4.	d	9.	a
5.	a	10.	c

Chapter 13

1.	c	6.	c
2.	d	7.	b
3.	d	8.	a
4.	a	9.	d
5.	b	10.	b

Chapter 14

1. a
2. d
3. a
4. c
5. a
6. c
7. b
8. a
9. e
10. c

Fianl Exam

1. a	22. c	43. f
2. e	23. c	44. c
3. f	24. b	45. b
4. e	25. a	46. e
5. b	26. d	47. d
6. c	27. b	48. e
7. e	28. c	49. e
8. e	29. a	50. e
9. e	30. d	51. b
10. b	31. c	52. c
11. e	32. a	53. c
12. e	33. b	54. d
13. c	34. c	55. b
14. b	35. a	56. d
15. b	36. d	57. b
16. d	37. f	58. a
17. b	38. a	59. d
18. c	39. b	60. b
19. a	40. b	61. c
20. d	41. a	62. d
21. a	42. c	63. a

64. c	77. a	90. d
65. c	78. c	91. e
66. b	79. b	92. b
67. d	80. c	93. a
68. a	81. e	94. d
69. c	82. a	95. b
70. a	83. c	96. c
71. c	84. d	97. b
72. a	85. c	98. c
73. c	86. b	99. a
74. b	87. a	100. e
75. a	88. c	
76. c	89. d	

APPENDIX

Common Structures

The following definitions will be helpful to you in understanding the structures of important biomolecules.

Acyl—a functional group with the formula $RC(=O)-$, with a double bond between the carbon and oxygen atoms, and a single bond between R and the carbon.

$$\underset{R}{\overset{O}{\underset{\|}{\bigwedge}}}$$

Aldehyde—a functional group which consists of a carbon atom bonded to a hydrogen atom and double-bonded to an oxygen atom (chemical formula $O=CH^-$). Also known as a formyl or methanoyl group.

Alkane—hydrocarbons consisting only of the elements carbon (C) and hydrogen (H), wherein these atoms are linked together exclusively by single bonds (i.e., they are saturated compounds) and without any cyclic structure. Also known as paraffins.

Alkene—unsaturated hydrocarbon containing at least one carbon-to-carbon double bond. Also known as olefin or olefine.

$$H \quad \quad H$$
$$\backslash \quad \quad /$$
$$C = C$$
$$/ \quad \quad \backslash$$
$$H \quad \quad H$$

Alkynes—hydrocarbons that have at least one triple bond between two carbon atoms.

$$H \quad \quad H$$
$$\backslash \quad \quad /$$
$$C \equiv C$$
$$/ \quad \quad \backslash$$
$$H \quad \quad H$$

Aliphatic—hydrocarbons in which carbon atoms are joined together in straight or branched chains. The simplest aliphatic compound is methane (CH_4).

Amines—a functional group of hydrocarbons that contain nitrogen.

Amide—a derivative of a carboxylic acid in which the hydroxyl group has been replaced by an amine or ammonia, represented by the formula: $R_1(CO)NR_2R_3$.

Apoprotein—protein component of a conjugated protein.

Aromatic—a hydrocarbon which incorporates one or more sets of six carbon atoms connected by delocalized electrons behaving as if they composed alternating single and double bonds. The configuration of six carbon atoms in aromatic compounds is known as a benzene ring.

Benzene ring—the simplest aromatic structure containing six carbon atoms and six hydrogen atoms connected by delocalized electrons distributed between covalent bonds.

Depictions of a benzene ring

Butyl—an unbranched alkane with four carbon atoms, $CH_3CH_2CH_2CH_3$. Also called *n*-butane.

n-Butane

Carbonyl—a functional group composed of a carbon atom double-bonded to an oxygen atom: C=O.

Carboxylic acids—acids containing a carboxyl group, with the formula C(=O)OH, usually written ⁻COOH or ⁻CO_2H.

Conjugated protein—a protein that functions in concert with another type of atom or molecule attached by covalent bonds or by weaker interactions. Also known as a holoprotein.

Delocalized electrons—orbital electrons that are not associated with a single atom or a covalent bond but are contained within an orbital that extends over several adjacent atoms.

Ester—acids in which a hydroxyl (⁻OH) group is replaced by an O⁻ group. The most common type of esters are carboxylic acid esters (R_1C(=O)OR_2).

Heme—a prosthetic group that consists of an iron atom contained in the center of a porphyrin. Also known as haem.

Hydroxyl—the functional group ⁻OH as part of a hydrocarbon. Organic molecules containing a hydroxyl group are called alcohols.

Imidazole—a class of ring structure hydrocarbons of the formula $C_3H_4N_2$. Acts as a base and also as a weak acid.

Examples of imidazoles

Indole—a bicyclic organic compound with a benzene ring fused to a pyrrole ring. Nitrogen electron is distributed so does not behave as a base.

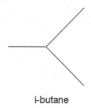

Examples of indoles

Isobutyl—a butyl with a branched chain. Also known as *i*-butane

i-butane

Keto acids—contain a ketone functional group and a carboxylic acid group.

Ketone—a functional group with a carbonyl group linked to two other carbon atoms, represented by the formula $R_1(CO)R_2$.

Methine—functional group with the formula CH^-.

Methyl—functional group with the formula CH_3.

Methylene—functional group with the formula CH_2^-.

Phenyl—functional group consisting of six carbon atoms arranged in a cyclic ring structure.

R—

Phenyl Group

Prosthetic group—nonprotein (nonamino acid) component of a conjugated protein. May be organic (many vitamins) or inorganic (such as the iron in hemoglobin).

Porphyrin—a macromolecule composed of four pyrrole subunits connected by methine bridges. Tend to be deeply colored.

Pyrrole—a cyclic compound containing nitrogen with the formula, C_4H_5N.

Depictions of Pyrrole Ring

Pyrrolidine ring—a cyclic amine with a five-membered ring containing four carbon atoms and one nitrogen atom represented by the formula C_4H_9N.

Saturated hydrocarbon—a chain of carbon atoms, each of which is connected to four different atoms or molecules, containing no double or triple bonds.

Thiols—a functional group containing a sulfur atom and a hydrogen atom (^-SH) Also known as a sulfhydryl group or a mercaptan.

Unsaturated hydrocarbon—a chain of carbons not all of which are bonded to four separate molecules, but, instead, contains double or triple bonds.

INDEX